Christian Dethloff

Modeling of Helium Bubble Nucleation and Growth in Neutron Irradiated RAFM Steels

**Schriftenreihe
des Instituts für Angewandte Materialien**

Band 6

Karlsruher Institut für Technologie (KIT)
Institut für Angewandte Materialien (IAM)

Eine Übersicht über alle bisher in dieser Schriftenreihe erschienenen Bände
finden Sie am Ende des Buches.

Modeling of Helium Bubble Nucleation and Growth in Neutron Irradiated RAFM Steels

by
Christian Dethloff

Dissertation, Karlsruher Institut für Technologie (KIT)
Fakultät für Maschinenbau
Tag der mündlichen Prüfung: 20. Juli 2012

Impressum

Karlsruher Institut für Technologie (KIT)
KIT Scientific Publishing
Straße am Forum 2
D-76131 Karlsruhe
www.ksp.kit.edu

KIT – Universität des Landes Baden-Württemberg und
nationales Forschungszentrum in der Helmholtz-Gemeinschaft

KIT Scientific Publishing 2012
Print on Demand

ISSN 2192-9963
ISBN 978-3-86644-901-5

Modeling of Helium Bubble Nucleation and Growth in Neutron Irradiated RAFM Steels

Zur Erlangung des akademischen Grades

Doktor der Ingenieurwissenschaften

der Fakultät für Maschinenbau
Karlsruher Institut für Technologie (KIT)

genehmigte

Dissertation

von

Dipl.-Ing. Christian Dethloff

aus Kassel

Tag der mündlichen Prüfung: 20.07.2012
Hauptreferent: Priv.-Doz. Dr. Jarir Aktaa
Korreferent: Prof. Dr. Robert Stieglitz

Parts of this thesis were published in the following articles:

- C. Dethloff, E. Gaganidze, V. V. Svetukhin, J. Aktaa
  Modeling of helium bubble nucleation and growth in neutron irradiated boron doped RAFM steels
  Journal of Nuclear Materials **426** (2012) 287–297

- E. Gaganidze, C. Petersen, E. Materna-Morris, C. Dethloff, O. J. Weiß, J. Aktaa, A. Povstyanko, A. Fedoseev, O. Makarov, V. Prokhorov
  Mechanical properties and TEM examination of RAFM steels irradiated up to 70 dpa in BOR-60
  Journal of Nuclear Materials **417** (2011) 93–98

- C. Dethloff, E. Gaganidze, V. V. Svetukhin, J. Aktaa
  Modeling of Helium Bubble Growth in 10B-doped RAFM Steels under Neutron Irradiation
  Poster presentation at *15th International Conference on Fusion Reactor Materials*, 2011, Charleston, SC, United States of America

- E. Gaganidze, C. Dethloff, O. J. Weiß, V. Svetukhin, M. Tikhonchev, J. Aktaa
  Modeling and TEM Investigation of Helium Bubble Growth in RAFM Steels under Neutron Irradiation
  Journal of ASTM International (2012), DOI:10.1520/STP103972

- C. Dethloff, E. Gaganidze, V. Svetukhin, M. Tikhonchev, O. Weiß, J. Aktaa
  Modeling of helium bubble growth in neutron irradiated boron doped RAFM steels
  Proceedings of *5th International Conference on Multiscale Materials Modeling*, 2010, Freiburg, Germany
  Fraunhofer Verlag, ISBN 978-3-8396-0166-2, p.675–678

- C. Dethloff, E. Gaganidze, V. Svetukhin, M. Tikhonchev, O. Weiß, J. Aktaa
  Modeling of helium bubble formation and growth in RAFM steels under neutron irradiation
  Proceedings of *1st International Conference on Materials for Energy*, 2010, Karlsruhe, Germany
  DECHEMA e.V., ISBN 978-3-89746-117-8, p.196–198

Abstract

Reduced Activation Ferritic/Martensitic (RAFM) steels are promising candidates for structural materials in future fusion technology. A large amount of helium will be generated by neutron irradiation induced transmutation during the lifetime of a fusion reactor. In addition to other irradiation defects, the transmuted helium is believed to strongly influence material hardening and embrittlement behavior.

A physically based model using Rate Theory is developed to describe nucleation and growth of helium bubbles in neutron irradiated RAFM steels. The model follows classical homogeneous nucleation and growth theory. The kinetic rate equations are set up on a distinct helium atom basis and are solved numerically by a self-written Fortran code. As a result, the time dependent evolution of the helium concentration and helium bubble size distributions are obtained.

The model is adapted to two different ^{10}B doped model alloys, ADS2 and ADS3, based on the composition of the RAFM steel EUROFER97. These steels were irradiated in the High Flux Reactor in Petten within the SPICE irradiation program and the BOR60 reactor in Dimitrovgrad within the ARBOR1 irradiation program. The basic model with specifically derived standard parameters for SPICE (ARBOR1) predict peak bubble diameters of 3.7 and 4.6 nm (6.9 and 7.9 nm) for ADS2 and ADS3, respectively. Values for specimens after SPICE irradiation agree well with preliminary Transmission Electron Microscopy (TEM) investigations, while the model overestimates bubble sizes for ARBOR1 when compared to quantitative TEM measurements. The effect of helium resolution from bubbles due to displacement cascade interaction is shown to be a potential solution, leading to smaller bubble sizes especially for ARBOR1 simulations.

Grain boundaries (GBs) and dislocations (DLs) acting as sinks for helium are implemented into the model. For the derived simulation parameters, GBs show a higher sink strength. A two-dimensional (2-D) variant of the model is developed to describe bubble nucleation and growth at GB sink sites. A comparison of simulated bulk and GB bubbles shows larger peak diameters in the latter case.

iv

Higher temperatures promote helium capturing by sinks and the following bubble formation and growth at these sites.

Helium induced hardening is calculated by applying the Dispersed Barrier Hardening (DBH) model to simulated cluster size distributions. Good agreement with experimental results from literature is achieved by assuming a linear superposition of helium induced hardening with hardening from other obstacle types.

The model is successfully applicable to conditions estimated for fusion allowing evaluation of helium bubble size distributions and expected hardening values.

Kurzzusammenfassung

Niedrig aktivierbare ferritisch/martensitische (*engl.:* RAFM) Stähle wurden speziell für den Einsatz als Strukturmaterial in zukünftigen Fusionsreaktoren entwickelt. Trotz ihrer, im Vergleich zu konventionellen rostfreien Stählen, verbesserten Widerstandsfähigkeit gegenüber Neutronenbestrahlung werden auch hier Strahlenschäden induziert. Neben erzeugten Frenkel–Punktdefekten (Leerstellen und Zwischengitteratome), die zur Bildung von Poren oder Versetzungsringen führen können, wird vor allem das durch Elementumwandlung gebildete Helium für eine Verschlechterung der mechanischen Eigenschaften dieser Stähle verantwortlich gemacht.

Diese Arbeit befasst sich mit der Entwicklung eines physikalisch basierten Modells, welches die Bildung und das Wachstum von Heliumblasen in neutronenbestrahlten Stählen beschreibt. Das Modell bedient sich der Raten–Theorie und folgt dem klassischen Ansatz der homogenen Keimbildung. Die aufgestellten Ratengleichungen beschreiben das Wachstum von Heliumblasen, indem die Wechselwirkungen zwischen Heliumblasen, gelöstem Helium in der Matrix und weiteren Gitterdefekten geometrisch und thermodynamisch betrachtet werden. Das gesamte Gleichungssystem wird numerisch mit Hilfe eines selbstentwickelten Fortran-Algorithmus gelöst. Die Berechnungen erlauben eine zeitabhängige Bestimmung sowohl der Heliumkonzentration in der Matrix als auch der Größenverteilung der Heliumblasen.

Das Modell wird verwendet, um die Entwicklung der Heliumblasen in den ^{10}B–dotierten Modellegierungen ADS2 und ADS3 zu beschreiben. Diese Legierungen basieren auf der Zusammensetzung des europäischen RAFM Stahls EUROFER97, wobei das zusätzlich dotierte Bor eine Erhöhung der Heliumproduktionsrate hin zu fusionsrelevanten Werten bewirkt. Proben aus diesen Materialien wurden bereits in den Bestrahlungsexperimenten SPICE und ARBOR1 bestrahlt, und Ergebnisse aus mikrostrukturellen und mechanischen Untersuchungen veröffentlicht. Diese Ergebnisse dienen nachfolgend zur Validierung der Simulationsergebnisse.

Die Simulationsparameter wurden speziell an die vorliegenden Bedingungen in den beiden Bestrahlungsexperimenten (z.B. Temperatur, Schädigungsrate, Bor-Helium-Umwandlungsrate) und bordotierten Stählen (Heliumgehalt) angepasst. Die Simulationen ergaben die höchste Blasenkonzentration für folgende Blasendurchmesser: Für SPICE wurden Werte von 3,7 nm (ADS2) und 4,6 nm (ADS3) berechnet, für ARBOR1 6,9 nm (ADS2) sowie 7,9 nm (ADS3). Ein Vergleich mit vorläufigen qualitativen Untersuchungen mittels Transmissionelektronenmikroskopie (TEM) zeigt eine gute Übereinstimmung der Simulation mit den Messungen für das SPICE–Bestrahlungsexperiment. Das Modell überbewertet jedoch die erreichten Blasendurchmesser für das ARBOR1–Experiment. Eine Modifizierung des Modells, bei der eine Wiederauflösung von Helium aus bestehenden Heliumblasen durch die auftreffende Neutronenstrahlung verursacht wird, bewirkt eine Verschiebung des maximalen Blasendurchmessers hin zu kleineren Werten. Dieser Effekt ist aufgrund der höheren Schädigungsrate in ARBOR1 besonders ausgeprägt und führt zu einer besseren Übereinstimmung mit experimentellen Ergebnissen.

Korngrenzen und Versetzungen werden als Senken für Helium in Betracht gezogen, eine Modellerweiterung beschreibt den Verlust von in der Matrix gelöstem Helium zu diesen Gitterdefekten. Auf Basis der verwendeten Simulationsparameter für SPICE und ARBOR1 zeigen Korngrenzen eine höhere Senkenstärke als Versetzungen. Da die Löslichkeit für Helium an den Korngrenzen beschränkt ist, wird eine zweidimensionale Modifikation des entwickelten Modells genutzt, um die Keimbildung und das Wachstum von Heliumblasen ebenfalls an Korngrenzen zu beschreiben. Ein Vergleich der ermittelten Blasengrößenverteilungen für das Volumen und an den Korngrenzen zeigt, dass das zweidimensionale Modell größere Blasen an den Korngrenzen vorhersagt. Höhere Temperaturen begünstigen dabei sowohl die Diffusion zu den Senken als auch das Blasenwachstum innerhalb der Korngrenzen.

Die simulierten Größenverteilungen der Heliumblasen werden verwendet, um die durch Helium verursachte Verfestigung der bordotierten Stähle abzuschätzen. Dazu wird das *Dispersed Barrier Hardening* (DBH) Modell verwendet, welches eine Erhöhung der Fließgrenze des Matrixmaterials auf die Behinderung von Versetzungsgleiten durch Hindernisse, in diesem Fall Heliumblasen, zurückführt. Eine gute Übereinstimmung mit experimentellen Untersuchungen zeigt sich unter der Annahme einer linearen Überlagerung der Verfestigung verursacht durch Helium und andere Gitterdefekte, z.B. Versetzungsringe.

Das Modell ist in der Lage, mikrostrukturelle Veränderungen unter Bestrahlungsbedingungen zu simulieren, wie sie in der Ersten Wand eines Fusionsreaktors erwartet werden. Im Vergleich zu den Bestrahlungsexperimenten SPICE und ARBOR1 werden dabei sowohl höhere Temperaturen als auch höhere Heliumproduktionsraten erreicht. Das hier entwickelte Modell kann dabei als Grundlage dienen, den negativen Einfluss von Helium auf die mechanischen Eigenschaften der RAFM Stähle abzuschätzen.

Acknowledgments

This doctoral thesis has originated from my work at the Institute of Applied Materials (IAM-WBM) of Karlsruhe Institute of Technology from July 2008 to December 2011. I would like to thank everyone who supported me in successfully completing my Ph.D. thesis.

I would like to thank Priv.-Doz. Dr. Jarir Aktaa for being the first examiner of my work. As head of department of Mechanics of Solids II he always offered helpful advice and support. My thanks go to Prof. Dr. Robert Stieglitz, who – quite on short notice – filled in and accepted to be the second examiner of this work. I would also like to thank Prof. Dr. Oliver Kraft for giving me the opportunity of obtaining my doctorate at his institute.

My special thanks go to Dr. Ermile Gaganidze. As my supervisor he showed great interest in my work, his ideas and our discussions provided important contributions to this thesis. Funded by Helmholtz Association and Russian Foundation for Basic Research within the framework of the joint research group HRJRG-13, I was given the opportunity to collaborate with Prof. Dr. Vyacheslav V. Svetukhin at the Ulyanovsk State University. He and his colleagues offered me a cordial welcome, and I am thankful for their willingness of sharing profound work-related knowledge, as well as for gaining new cultural and personal experience.

I would like to thank all Ph.D. students and colleagues from IAM-WBM. Besides offering professional help and support they were responsible for creating a good working atmosphere. Beyond that frequently organized off-work activities introduced me to new hobbies and friends.

Finally, I would like to thank my parents and my brother for their interest and support.

Karlsruhe, July 2012   Christian Dethloff

Contents

Nomenclature

GB	grain boundary. iii, 65, 68–70, 103, 105, 108, 110, 122, 127–129, 134, 135
GIF	Gatan Image Filter. 47
HFIR	High Flux Isotope Reactor. 54
HFR	High Flux Reactor. 17, 18, 20, 39, 113, 123
IFMIF	International Fusion Materials Irradiation Facility. 18
ITER	International Thermonuclear Experimental Reactor. 2, 8, 9
JANNUS	Joint Accelerators for Nanosciences and NUclear Simulation. 52
JET	Joint European Torus. 2
KIT	Karlsruhe Institute of Technology. 39, 47
kMC	kinetic Monte Carlo. 24, 34, 115
LCF	Low Cycle Fatigue. 18
LR	long range. 38
MC	Monte Carlo. 24, 57
MD	Molecular Dynamics. 23, 24, 34, 35, 37, 55, 57, 58, 62, 66, 67, 70, 106, 113, 123, 124, 134
MS	Molecular Static. 35, 58
NaN	Not a Number. 73
ODS	oxide dispersion strengthened. 15, 16, 38
PKA	primary knock-on atom. 10, 54, 55, 66, 67, 123, 124
ppm	parts per million. 5, 7
RAFM	Reduced Activation Ferritic/Martensitic. iii, 2, 3, 8–12, 15–20, 38, 39, 54, 91, 112, 113, 132, 133
RED	Radiation Enhanced Diffusion. 10, 35, 92

RSS	root sum square. 38, 70, 110, 125, 126, 135
SD	size distribution. 87, 89, 90
SEM	Scanning Electron Microscope. 45
SEM	Scanning Electron Microscopy. 17, 47, 49
SIA	Self Interstitial Atom. 10, 12, 25, 30, 33, 35
SINQ	Swiss spallation neutron source. 16, 21
SR	short range. 38
SSC-RIAR	State Scientific Center - Research Institute of Atomic Reactors. 17
STEM	Scanning Transmission Electron Microscopy. 47
TEM	Transmission Electron Microscopy. iii, 10, 11, 15, 16, 20, 45, 47, 49, 50, 52, 54, 68, 112, 113, 115, 118–122, 128, 133, 135
USE	Upper Shelf Energy. 15, 42
UTS	Ultimate Tensile Strength. 42

Notation List

α_j .... barrier strength of obstacle j

Δt .... time step

$\Delta\sigma_{\text{tot}}$ .... total hardening

$\Delta\sigma_j$ .... hardening caused by defect type j

γ_{SF} .... macroscopic surface energy (of iron)

γ_i .... surface energy per cluster atom

$\langle d \rangle$ .... mean bubble diameter

μ .... shear modulus

Ω .... atomic volume

ρ_{LD} .... dislocation density

σ_{ys} .... yield stress

θ_i .... internal degrees of freedom for defect type i

A .... atomic mass

a_{Fe} .... iron lattice constant

A_{gr} .... grain surface

A_i .... effective surface

b .... equivalent monomer radius

b_{disl} .... Burgers vector of gliding dislocation

C^{He} .... cumulative helium concentration

C_1 .... concentration of monomers

C_1^{eq} .... helium solubility

C_{10B}^{max} .... ^{10}B content

C_i .... interstitial concentration in the bulk volume

C_i^0 .... thermodynamic equilibrium interstitial concentration

C_i^{V} .... interstitial concentration at the void surface

C_{loop} .... interstitial loop concentration

C_{v} .... vacancy concentration in the bulk volume

C_{v}^{0} .... thermodynamic equilibrium vacancy concentration

C_{v}^{V} .... vacancy concentration at the void surface

C_i .... concentration of cluster with size i

C_i^{aff} .... affected concentration of cluster with size i by cascade induced helium resolution

C_i^{eq} .... equilibrium concentration of cluster with size i

$D_{\text{He}}^{\text{eff}}$ .... effective helium diffusivity

D_{i} .... diffusivity of single interstitials

D_{v} .... diffusivity of single vacancies

d_j .... size of defect type j

E_{m} .... migration energy

E_{PKA} .... kinetic energy of primary knock-on atom

E_i .... attachment barrier

err .... allowed change of concentration per time step

F_{v} .... shape factor

G .... Gibbs free energy

G_{dpa} .... damage rate

G_{dpa}^{0} .... characteristic damage constant for ^{10}B transmutation

G_{He} .... helium generation rate

G_{matrix} .... helium transmutation rate of matrix elements

G_{surf} .... surface parameter

G_{vol} .... volume parameter

g_i .... monomer emission rate of clusters with size i

G_i^{f} .... Gibbs free energy of formation of cluster with size i

G_i^{n} .... nucleation energy of cluster with size i

h .... Planck constant

i .... cluster / bubble size

i_{calc} .... largest bubble size used in the actual calculation step

i_{max} .... maximum considered bubble size

j .... defect type

k_{sinks}^2 .... total sink strength
k_j^2 .... sink strength of defect j
k_{B} .... Boltzmann constant
k_i .... monomer capture rate of clusters with size i
$k_{i,\text{diff}}$ .... diffusion governed capture rate
$k_{i,\text{react}}$ .... reaction governed capture rate
L_{sinks} .... helium loss rate to sinks
l_j .... mutual spacing of obstacles of type j
M .... monomer
m .... helium mass
M_{T} .... Taylor factor
M_i .... cluster of size i
n_{He} .... number of helium atoms in bubble
$n_{\text{He}}^{\star}$ .... critical number of helium atoms in bubble
N_j .... density of defect type j
p .... bubble pressure
R .... radius
R_{cr} .... critical radius
$R_{\text{cr}}^{\star}$ .... critical cluster radius
$R_{\text{cr}}^{\text{bub}}$ .... critical bubble radius
$R_{\text{cr}}^{\text{void}}$ .... critical void radius
R_{eff} .... effective bubble/void radius
R_{grain} .... grain radius
R_{loop} .... interstitial loop radius
R_i .... cluster radius
S_{F} .... formation entropy
S_{v} .... effective vacancy supersaturation
T .... temperature
t .... time
tol_{abs} .... absolute tolerance
tol_{rel} .... relative tolerance
V^{V} .... cavity volume

V_1 .... monomer volume

V_{gr} .... grain volume

V_i .... volume of cluster with size i

X .... number of possible positions of helium atoms per number of unit cell atoms

X_{corr} .... correlation factor

x_{He} .... helium-to-vacancy ratio

x_i .... atomic fractions of different defect types i

Z .... atomic number

Z_{He} .... bias factor of dislocation sink strength

D .... deuterium

He_1 .... interstitial helium atom

He_1V_1 .... substitutional helium atom

He_1V_2 .... helium-divacancy cluster

n .... neutron

n' .... neutron with altered kinetic energy compared to n

p .... proton

T .... tritium

V .... vacancy

1. Introduction

Ensuring the electrical power supply for following generations will be one of the main crucial tasks in the near future. Even though the energy situation in the industrialized countries seems quite satisfying at the moment, we have to deal with that problem right now to find an acceptable solution within the next decades. Fossil fuels like coal, crude oil and natural gas are limited and resources will probably draw to a close within this century [1], at least the exploitation of the raw materials will become more and more extensive as well as expensive. Burning fossil fuels to create thermal energy leads to a high production of the greenhouse gas carbon dioxide, whose emission is strictly monitored and limited within the European Union. Renewable energy sources using solar, wind and water energy seem promising candidates for future electrical power supply concerning CO_2 emission, but the low capability and especially the non-continuous energy production due to environmental changes limits their application as a technology, which can be used to ensure the base load of energy generation. Unlike renewable energy sources nuclear power plants are widely used to create precisely the fore mentioned basic load energy. Identified resources of uranium will last for a hundred years based on data of 2008, reprocessing and recycling used uranium increases that time period to a thousand years [1]. Nevertheless, the general acceptance of nuclear fission energy is quite low due to risks of radiation accidents and the unresolved problem of nuclear waste disposal.

Not only is it mandatory to look for power alternatives to replace the energy supply in times of diminishing raw materials, but also the highly escalating energy needs of newly industrialized countries like China, India and further countries in Southeast Asia with their billions of people have to be fulfilled. From 2010 to 2035, the world's electricity demand is expected to increase by around 80%, which requires an increase in energy consumption of fuels between 21% and 47% depending on the considered energy scenario [2].

It is obvious that the mentioned reasons demand alternative ways of energy production in the future. Besides increasing efforts in new design ideas based on renewable energy sources, e.g. concentrated solar power or solar updraft tower power plants, special attention is paid worldwide to the technological development of fusion as a usable energy source. The idea is far from being new: it has been more than half a century since scientists recognized the potential of thermo–nuclear fusion for power generation, and started to invent ways of bringing the process, which fuels the sun for billions of years, down to earth. That does not only sound ambitious, it requires sophisticated technologies, which have been examined and developed to some extent during the last decades, to ignite and stabilize a fusion plasma on earth. Research was thereby not exclusively organized on national authorities, but soon an international cooperation was arranged which coordinates the efforts in each member country. Up to now many fusion experimental facilities, like e.g. ASDEX in Germany and JET in the UK, have been built, to prepare the physical base for International Thermonuclear Experimental Reactor (ITER) and DEMO and to develop and study the necessary technologies [3]. The actual task is to build ITER, where for the first time all crucial technologies, which up to now have been tested only seperately, will be joined together. Although ITER will still be a small scale version of a future fusion power plant with some parts not designed for continuous running, e.g. testing tritium breeding modules for the First Wall (FW), it is supposed to demonstrate for the first time a positive energy balance while operating [4].

One key aspect of harnessing fusion energy as an energy source is the engineering of new materials. Fusion technology makes high demands on functional and structural materials because the fusion environment differs from known conditions, e.g. in nuclear fission power plants. While ITER's FW of the reactor will be built out of austenitic stainless steel, it is not the materials first choice for future fusion reactors. Fusion power plants will operate for years, leading to high thermal and neutron loads on the reactor containment. To avoid extensive radioactive waste disposal, activation of the used materials has to be minimized. Therefore a new class of steels has been developed, which shows low activation and high radiation resistance: these Reduced Activation Ferritic/Martensitic (RAFM) steels, like the European EUROFER97 [5], are first candidate materials for the FW of the reactor.

Available since 1999, EUROFER97 was subjected to many scientific studies, determining materials and mechanical properties as well as behavior under irradiation. For this purpose irradiation programs were established where RAFM model alloys were exposed to neutron irradiation from conventional experimental fission reactors. While those ferritic steels show better radiation resistance than austenitic steels, low temperature neutron irradiation causes hardening and embrittlement, which is examined with great anxiety. Helium generated by transmutation is expected to have a great influence on the mechanical properties of RAFM steels under fusion conditions. Unfortunately, irradiation programs show strong limitations: although the applied overall damage is comparable to expectations in fusion power plants, neutron flux and especially neutron spectra differ significantly between fission and fusion. This fact leads to different defect production rates, in particular helium generation by transmutation: a constant helium production rate at high helium levels as fusion will provide can not be achieved.

Obviously, irradiation experiments can not sufficiently reproduce conditions under fusion. Therefore, due to advanced computational potential, various simulation methods are used to gain insight into the reasons and consequences of materials' irradiation behavior. Several types of simulations allow description of the behavior from interactions on the atomistic level at small time scales, through dislocation movement for plastic deformation, till years of defect cluster growth on a microscopic scale.

In this work, simulations aim at precisely describing existing irradiation experiments on a physical basis. Rate theory is used to describe helium nucleation and growth for these experiments. The model can be verified by existing microstructural data on unirradiated and irradiated specimens of RAFM steel EUROFER97. Mechanical influences of helium bubbles can be assessed by comparing simulations with results from mechanical investigations. Therefore it is possible, after building up a model for describing the irradiation experiments, to modify the model by adapting relevant parameters to expected conditions under fusion. Conclusions can thus be drawn on expected lifetimes of structural steels under fusion environments.

This work is structured as follows: Chapter 2 supplies background information on fusion, irradiation experiments and consequences of neutron irradiation. Chapter 3 summarizes state-of-the-art research, providing specific results relevant for this work. In Chapter 4 the developed basic model is derived together with its implemented modifications, while Chapter 5 presents the numerical approach for solving the rate equations. Chapter 6 shows simulation results, which are discussed in Chapter 7. Chapter 8 summarizes the achievements of this work and gives an outlook.

2. Background

2.1. Principles of fusion

The fusion process, which takes place in our sun's core, consists of proton-proton chain reactions resulting in the creation of helium ions. The mass-energy conversion therein yields an energy rate of sun's surface emission of more than $63\,\mathrm{MJm^{-2}s^{-1}}$ [6]. While this process works well for stars, it can not be used in a fusion reactor on earth. The reasons for this, the main principles, and the preferential fusion reaction will be explained in this section.

In thermonuclear reactions, energy will be released if the reaction products take a lower energy state and thus are more stable than the starting nuclei. Fig. 2.1a shows the binding energy per nucleon for several isotopes [7], where the highest values are possessed by elements close to ^{56}Fe. Therefore an exothermic core reaction can be achieved by splitting heavier nuclei into lighter ones or combining lighter cores into heavier ones. While the former reaction is applied in nuclear fission reactors using e.g. uranium, the combination of light hydrogen isotopes is energetically highly profitable. This is also due to the fact that light isotopes show a lower repelling force when compared to heavier ones, thus reducing the necessary temperature for the fusion process. As mentioned earlier, the sun runs a proton-proton chain reaction to create helium. Although these reactions show the highest energy release, the process possesses a low reaction cross section (Fig. 2.1b) when compared to other reactions [8]. Therefore, taking into account energy release and reaction probability, the combination of the hydrogen isotopes deuterium (^{2}H, or D) and tritium (^{3}H, or T) in a fusion process creating ^{4}He is more efficient.

Besides fusion efficiency, availability of the hydrogen isotopes used as fusion fuel is important. Deuterium is a stable hydrogen isotope, which naturally occurs with a fraction of 150 atomic parts per million (ppm) [8]. ^{2}H can be extracted by electrolysis from sea water [9], which provides an almost infinite resource

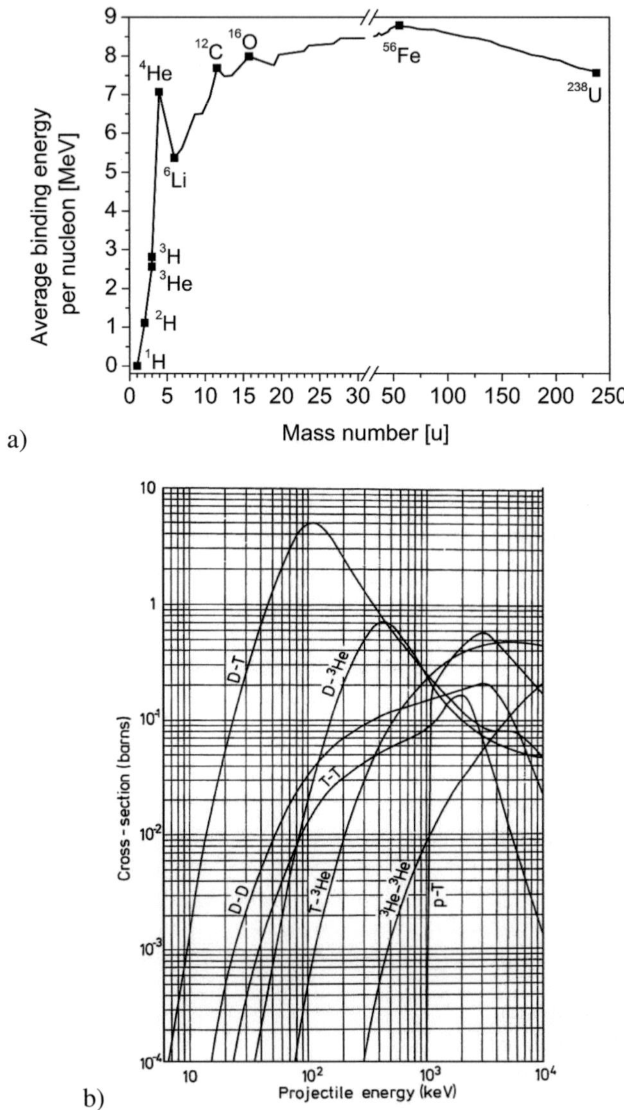

Figure 2.1.: a) Average binding energy per nucleon for several isotopes [7]. High binding energies correlate with stable elements, which is why energy can be released by joining light elements like hydrogen (fusion) or splitting up heavy elements like uranium (fission). b) Reaction cross sections for possible fusion candidate isotopes [8]. For a low projectile energy, i.e. temperature, D-T fusion owns the highest reaction probability.

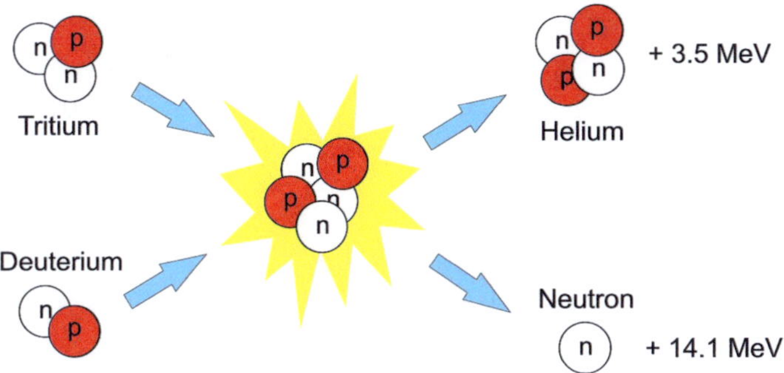

Figure 2.2.: Fusion reaction: the charged heavy hydrogen isotopes deuterium and tritium collide, forming helium (alpha particle) and a free neutron. The neutron holds 80% of the kinetic energy, which is transformed into thermal energy in the First Wall of the fusion reactor.

of deuterium. Tritium, however, is radioactive and naturally not accessible. Nevertheless, thermal neutrons created by the fusion process can be used for in situ production of ^{3}H out of ^{6}Li (natural abundance 7.5%) by the reaction

$$^6\text{Li} + {}^1\text{n} \longrightarrow {}^4\text{He} + {}^3\text{H} + 4.8\,\text{MeV}, \tag{2.1}$$

while the reaction of ^{7}Li (92.5% of lithium's isotopes) to tritium is also possible but less likely. The mass fraction of lithium in earth's geosphere is described to lie between 20 and 60 ppm [10, 11]. Taking into account that the amount of lithium contained in one laptop battery would provide enough fusion energy for a human's lifetime electricity needs [12], availability seems to be guaranteed.

The fusion reaction is shown in Fig. 2.2. To build helium, the positively charged nuclei have to approach so close that they overcome the repelling force of the coulomb energy. Inside stars, enormous pressures and high temperatures up to millions of degrees support the fusion process. Because no material would withstand such high temperatures, the fusion reaction has to be confined contact-free. Due to the high temperatures the reaction gases exist as a plasma of charged nuclei, which can be contained by magnetic fields and thermally insulated by an ultra-high vacuum. Although mandatory, vacuum lowers the density of the reacting isotopes, thus empeding a fusion reaction. Therefore, within a fusion reactor vessel, reacting nuclei have to be accelerated by even higher temperatures (around 100 million degrees) than inside stars, so that they hit each other with

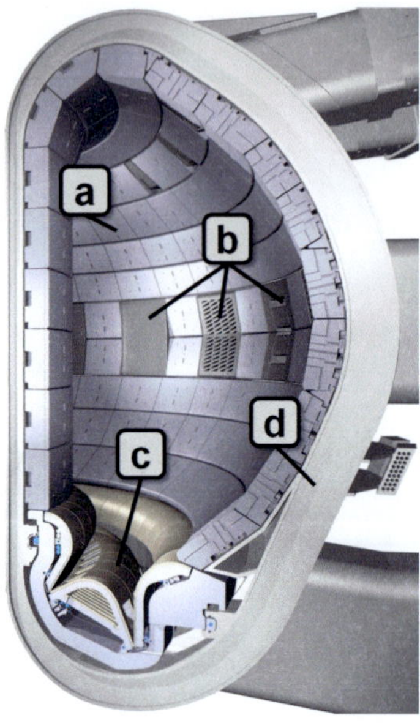

Figure 2.3: Cut-away of the ITER vacuum vessel [4]. **a** – Blanket Modules, **b** – access openings for heating systems, diagnostics, remote handling, **c** – divertor, **d** – vacuum vessel.

high kinetic energy to create a joint nucleus. When forming ^{4}He, a thermal neutron with a kinetic energy of 14.1 MeV is ejected. While the kinetic energy of the helium nucleus is used to heat the plasma and keep the fusion reaction in progress, the uncharged neutron escapes the magnetic containment and is moderated in the First Wall (FW) of the reactor to create thermal energy.

There exist two different design concepts of reactor geometry which are considered for future fusion power plants: *Stellarator* and *Tokamak*. Both confine the hot plasma by magnetic fields within a reactor pressure vessel under ultra-high vacuum. Since International Thermonuclear Experimental Reactor (ITER) will be based on the tokamak design, the key issues and technologies will be explained using this example. The cut-away of the toroidal vacuum vessel of ITER is shown in Fig. 2.3. The FW next to the plasma consists of Blanket Modules (BMs) (labelled *a* in Fig. 2.3), which serve the following purposes: BMs contain beryllium, which is used as a neutron multiplier to enhance efficiency. Alternating pebble beds of Be and Li are used to breed tritium according to Eq. (2.1). While in ITER, this function is only checked out for chosen Test-BMs built from different Reduced Activation Ferritic/Martensitic (RAFM) steels, e.g. EUROFER97 and

F82H-mod, breeding blankets are mandatory for future DEMO and fusion power plants. Through collisions impacting neutrons are moderated in the BMs, the generated heat is conducted by a coolant to a heat exchange circuit, which ends up in a turbine to generate electric energy. Tritium is extracted out of the (Test-)BMs, and used to fuel the fusion process by introducing it into the burning plasma. Before the fusion process can occur, the reaction gases have to be heated until they become a plasma with a temperature high enough for ignition. In ITER different external heating systems will be used: neutral beam injection and two types of high-frequency electromagnetic waves, which will be introduced into the vacuum vessel through special access openings (*b*). Maintenance after assembling is highly constrained, but further ports exist for diagnostics and remote handling (also symbolized by *b*). The role of the divertor (*c*) is to clean the plasma from exhaust gases (especially helium) and impurities, e.g. due to abrasion from plasma-FW interaction. The divertor has to withstand the highest thermal load of the vacuum vessel (*d*), and will therefore be coated by tungsten tiles. While for ITER the FW and Blanket will be constructed out of conventional stainless steel, in future fusion reactors most likely RAFM steels (see Section 3.4) will be used.

Gigantic magnets surround the vacuum vessel to keep the fusion plasma in shape. High temperature superconductors are necessary to achieve magnetic fields up to 12 Tesla without energy dissipation due to ohmic resistance. The whole construction is enclosed by a cryostat, providing thermal insulation against surroundings to keep the superconducting field coils below their critical temperature.

All technologies mentioned are required to fullfill the scientific goal of ITER: its produced power shall be ten times higher than the power it consumes. In 2019, when the production schedule projects the ignition of the first plasma [4], ITER has to demonstrate its feasibility and potential.

2.2. Effects of neutron irradiation on structural steels

High energy particle irradiation, especially neutron irradiation, causes severe damage in exposed materials. Collisions of impacting neutrons with matrix atoms lead to the creation of displacement cascades, where the kinetic energy

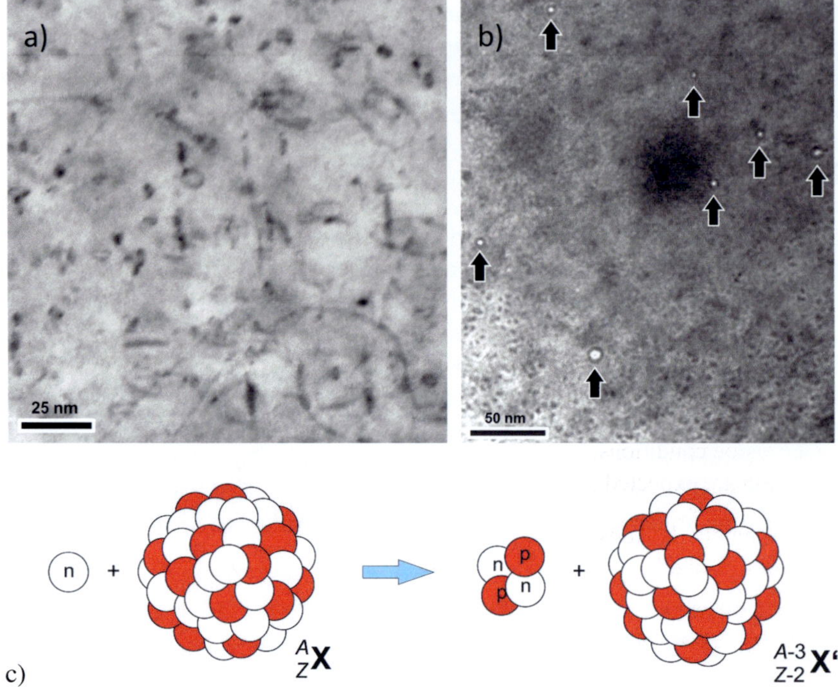

Figure 2.4.: Radiation induced defects. TEM investigation on EUROFER97 specimen after irradiation to 15 dpa at 330 °C showing a) dislocation loops and line dislocations [13] and b) voids [14]. In c), element **X** is transmuted by an impacting neutron to element **X'** producing a helium ion.

is transferred to a primary knock-on atom (PKA) which itself cause further collisions. Within this process Frenkel pair defects, i.e. a lattice atom moves to an interstitial lattice site and thus forms a Self Interstitial Atom (SIA) and a vacancy, are formed. Defects can diffuse through the matrix supported by Radiation Enhanced Diffusion (RED) and agglomerate: SIAs form clusters, which become discoidal dislocation loops due to energy minimization after reaching an appropriate size, while vacancies may form voids. Fig. 2.4a shows a Transmission Electron Microscopy (TEM) image of the microstructure of RAFM steel EUROFER97 after irradiation to a dose of 15 dpa at 330 °C [13]: one can observe dislocation lines and elliptical dislocation loops; also visible are black dots, which are either small defect clusters or dislocation loops, which can not be resolved by TEM due to their small size. In Fig. 2.4b voids up to sizes of 6 nm are observable [14]. They show the typical contrast with bright spot and dark fringe, which inverts when the focus of the electron beam is moved from under- to overfocus condition.

Neutron irradiation leads to transmutation reactions in the surrounding material (in our case steel). Particular attention is paid to (n,α)-reactions, which produce e.g. helium out of matrix elements. Fig. 2.4c describes the reaction of the thermal neutron from the fusion reaction with an element $\mathbf{X}$, which leads to the creation of an alpha particle (helium ion) and element $\mathbf{X'}$, with a reduced atomic mass A and atomic number Z. Solubility of helium in steel is very low, thus helium transmutation yields creation of helium bubbles in the matrix. While helium bubbles look identical to the voids in Fig. 2.4b, their density will be much higher under fusion conditions, where helium amounts of up to 450 appm/year of reactor operation are expected [15] (see also Tab. 2.2).

Unfortunately, besides helium transmutation, neutrons are responsible for activation of materials. In addition to (n,α)-reactions shown in Fig. 2.4c, many other interactions may take place: (n,n'), (n,p), (n,2n) etc., where an impacting neutron causes the emission of a neutron with different kinetic energy, a proton or two neutrons, respectively. Reactions thereby not only yield stable nuclides $\mathbf{X'}$, but also radioactive isotopes are created. Exemplarily, Fig. 2.5 shows the decay of the contact dose rate of RAFM steel EUROFER97 in the FW of the fusion DEMO reactor after 2.3 years of steady irradiation, derived by activation calculations [16]. Radioactive nuclides, which stem from alloying elements or impurities, are responsible for high and persistent activation, even though their concentration is low (see EUROFER97 composition in Tab. 3.1). While ^{54}Mn and ^{60}Co are responsible for high dose rates for a time period of less than 100 years, especially ^{94}Nb prevents radiation from reaching the hands-on contact dose rate limit of 10 μSv/h. Nevertheless, RAFM steels – as the name implies – show little activation when compared to conventional stainless steels (see also Section 3.4), whose decay time needs up to millions of years until the contact dose rate drops below the recycling limit.

Microstructural defects and impurities are responsible for changes in the mechanical properties. Effects are well known for face-centered cubic (fcc) austenitic stainless steels which have been thoroughly investigated and widely been used for nuclear applications. Generation and clustering of vacancies yield voids, which

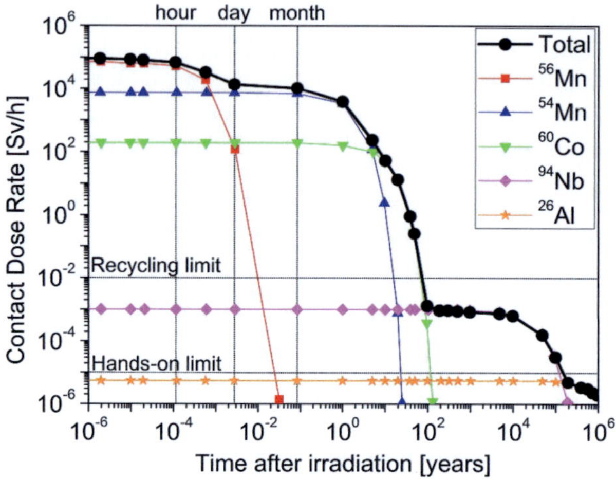

Figure 2.5.: Activation calculations for EUROFER97 showing the decay of the contact dose rate in the First Wall of a fusion DEMO reactor after 2.3 years of irradiation [16].

enlarge the volume of the materials and thus lead to swelling under irradiation. Fig. 2.6 shows swelling behavior for austenitic steel claddings under Phénix irradiation conditions ($T > 350\,°C$) [17] compared to ferritic/martensitic (F/M) steels [18]. Austenitic steels suffer high swelling, which is an important issue and can easily account for a volume increase of several percent after 100 dpa (see also [19]). F/M steels, on the other hand, show only slight swelling below 1% up to very high damage doses of 155 dpa. This radiation resistance is the reason why F/M steels were chosen as basis for further research and the development of RAFM steels. Swelling behavior, which has been thoroughly investigated for austenitic and low alloy steels [20, 21], depends strongly on temperature. At low temperatures, where both SIA and vacancy diffusion is slow, vacancies recombine with their Frenkel defect counterparts directly after creation. When temperature rises, diffusivity of SIAs increases faster than vacancy diffusivity. SIAs can easily diffuse through the matrix, and are trapped at microstructural sinks, e.g. grain boundaries and dislocations. This leads to forming of dislocation loops, while vacancies remain in the matrix, also starting to form voids and therefore increasing sample volume. At higher temperatures on the other hand, vacancies as well as SIAs have high diffusion coefficients, which leads to their mutual

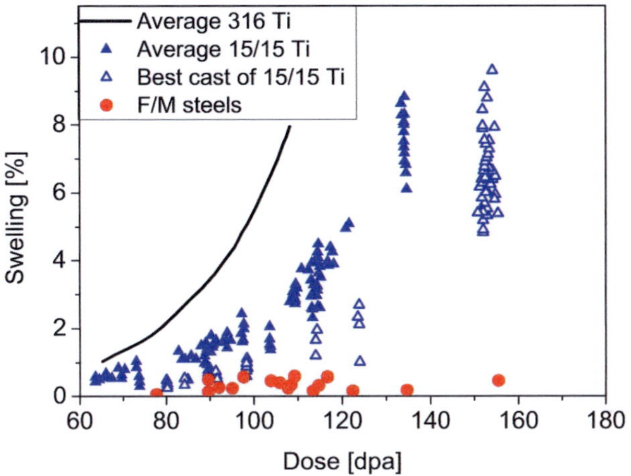

Figure 2.6.: Swelling of several austenitic steel claddings under irradiation ($T >$ 350°C) in Phénix fast breeder reactor [17] with comparison to ferritic/martensitic (F/M) steels [18].

trapping and annihilation at sinks before clustering occurs. It is quite evident that a great amount of pores has a deteriorating effect on the mechanical properties like strength and fracture toughness of the considered structural materials.

Dislocation loops are significantly involved in hardening of irradiated materials. By definition, hardening corresponds to an increase in yield strength. Dislocation loops, besides other microstructural defects like precipitates, cause stress fields which are responsible for impeding dislocation glide. Higher external stresses are necessary to activate other glide systems to bypass the obstacles, also yielding a loss of ductility due to a reduced number of freely moving dislocations. In Fig. 2.7a engineering stress-strain curves from tensile tests are shown for EUROFER97 WB [22]. In the unirradiated condition, yield strength is about 500 MPa at a test temperature of 300 °C, reaching a total strain at fracture of 20%. Irradiation at 300 °C to a dose of 15 dpa increases yield strength to a value of 900 MPa, but leads to a reduced total strain of about 10%. Other irradiation and test temperatures result in a lower yield strength, and hence the deteriorating effect of the neutron irradiation on hardening has a maximum at low temperatures around 300 °C.

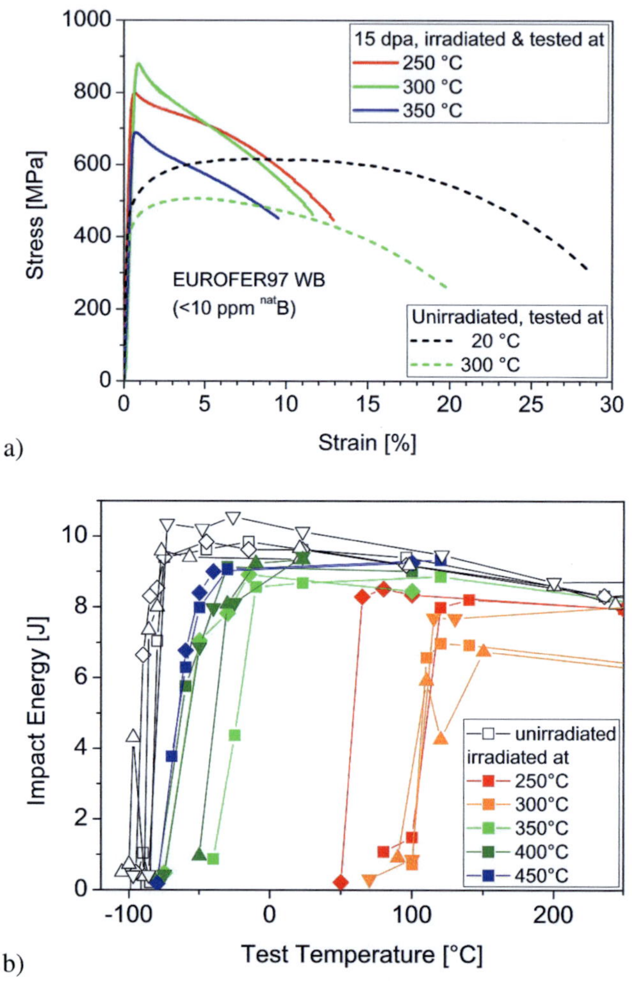

Figure 2.7.: a) Tensile tests of EUROFER97 WB specimens before and after irradiation to a dose of 15 dpa at different temperatures. An increase in yield strength together with a reduced total strain at fracture can be observed after irradiation [22]. The largest influence of irradiation was shown to be at an irradiation temperature of 300 °C. b) Embrittlement of RAFM steels EUROFER97 ANL (□), EUROFER97 WB (◇), F82H-mod (△), OPTIFER-Ia (▽) after irradiation to an average dose of 16.3 dpa [23]. At irradiation temperatures below 300 °C a strong shift of the DBTT towards higher temperatures is observed, while at irradiation temperatures above 350 °C embrittlement is less severe.

Materials with body-centered cubic (bcc) lattice structure typically show a transition in fracture behavior with rising temperature from predominant brittle to ductile failure. The effect is due to temperature dependent activation of glide systems, and is expressed by means of the Ductile-to-Brittle-Transition Temperature (DBTT). The DBTT is typically determined by impact test measurements [23], where fracture (impact) energy is measured by breaking a notched specimen with the help of an impacting pendulum hammer with specific velocity. Fig. 2.7b illustrates the effect of embrittlement, i.e. the shift of DBTT to higher temperatures under irradiation, taking into account unirradiated and irradiated specimens of different RAFM steels EUROFER97 (with name affixes ANL and WB depending on the pre-experimental heat treatment, see Tab. 3.2 in Section 3.4), F82H-mod and OPTIFER-Ia [23]. All steels show a similar embrittlement behavior. For EUROFER97 ANL and WB in the unirradiated state, DBTT lies at -81 °C and -91 °C [24], respectively. After irradiation at temperatures between 250 and 300 °C to 16.3 dpa, DBTT is strongly increased and shifted by values between 150 and 190 °C. For irradiation temperatures above 350 °C, a lower embrittlement is observed which is only slightly increased in comparison to the unirradiated state. The high energy level, called Upper Shelf Energy (USE), is associated with the ability to deform plastically, and can be compared to other materials investigated under the same testing conditions, i.e. mainly hammer impact velocity but also specimen geometry and size. Under low temperature neutron irradiation, DBTT is strongly shifted to higher temperatures with a substantial decrease of USE.

2.3. Helium formation

Microstructural investigations by TEM had been performed on different steels under different irradiation conditions in literature. Helium bubbles were observed for austenitic [25] and martensitic steels [26] after α-implantation, oxide dispersion strengthened (ODS) reduced activation ferritic steels after dual-beam injection with α and iron ions [27], RAFM steel EUROFER97 after α-implantation through nickel coatings [28] and boron doped EUROFER97 based RAFM model alloys after thermal neutron irradiation [29].

Conclusions on microstructural bubble observations are not straightforward. In summary, small amounts of produced helium (< 10 appm) are rarely observed in terms of bubbles, mainly due to the limited TEM resolution where bubbles with diameters of less than 1 nm can hardly be resolved. An increasing microstructural complexity, i.e. a high number density of lattice defects like dislocations, leads to void supression in the investigated steels. Furthermore, ODS particles reduce observed bubble sizes by providing a high number density of nucleation centers for helium bubbles, thus producing many sub-nanometer sized bubbles which for the most part are invisible for TEM.

Irradiation temperature strongly influences microstructural evolution. Specimens of boron doped EUROFER97 based model alloy ADS3 were studied after low temperature irradiation at 250 °C in SPICE (13.6 dpa) [29, 30] and at 335 °C in ARBOR1 (22.4 dpa) [31] irradiation experiments. Evaluation of the micrographs showed homogeneously distributed helium bubbles throughout the matrix, mostly with sizes of less than 5 nm for both experiments. By contrast, after irradiation at higher temperatures of 450 °C to 18.1 dpa in SPICE [29, Fig.7], most bubbles are located at lattice defects like dislocations and grain boundaries.

The main interest of investigating helium bubbles in structural steels is of course their influence on mechanical properties and therefore the deteriorating effects on durability under irradiation. In [32], mechanical investigations by tensile and impact tests on several structural F/M and austenitic steels are summarized, which were irradiated at Swiss spallation neutron source (SINQ). An examination of helium effects under irradiation separated from other defect influences is quite sophisticated, because helium production rates often are directly related to damage rates. Irradiation thus leads to an increase in helium, but also other defect densities are increased by interaction of displacement cascades with the host lattice. Nevertheless, taking the example of the RAFM steel F82H and EUROFER97 studied in [32], hardening and embrittlement was observed and to some extent attributed to helium bubbles. In the case of embrittlement, a comparison of EUROFER97 irradiated at SINQ and by fast neutrons is given. Because under neutron irradiation helium production is very low, differences in embrittlement for the same damage dose can be mainly attributed to the generated helium under SINQ irradiation: after $\sim$9 dpa the difference in DBTT between both irradiation conditions yield 100 °C, while after $\sim$18 dpa it was estimated to be 400 °C, corresponding to helium concentrations of 600 and 1400 appm, respectively.

In [33] helium effects on hardening and embrittlement were pointed out through results on RAFM steel F82H after neutron irradiation to 2.4 dpa at 250 °C. Using boron doping technique, helium production was adjusted and could be studied separately from other defect influences. The generated helium amount varying between 5 and 330 appm yield progressive helium induced hardening and embrittlement up to 100 MPa and 115 °C, respectively.

In [34] microstructural and mechanical investigations were performed on 9Cr-1Mo martensitic steel T91. Impact specimens (KLST) were irradiated at the notch region by a 34 MeV ^{3}He beam, implanting helium (0.25 at%) homogeneously at 250 °C up to a depth of 240 μm. Three-point bending tests were performed at room temperature, and the fracture surface was studied by Scanning Electron Microscopy (SEM). It was shown that directly below the notch, fully brittle fracture appeared with intergranular fracture and cleavage. This depth exactly corresponds to the helium implantation depth. Afterwards, in the zone without helium, exclusively ductile fracture was observed. It was concluded that helium is responsible for decreasing the critical stress for intergranular fracture, as well as the fracture toughness.

2.4. Irradiation facilities / programs

For further application as structural material in FW and breeding BMs of DEMO and upcoming commercial fusion reactors, mechanical behavior of RAFM steels under (neutron) irradiation is indispensable to be known and understood. While several experimental fission reactors for specimen testing exist worldwide, two comprehensive high dose irradiation programs for the European EUROFER97 were performed: HFR Phase IIb (SPICE) [23, 30] and ARBOR [35]. The experiment SPICE took place at the High Flux Reactor (HFR) in Petten (Netherlands) [36], dealing with the influence of irradiation temperature on microstructure and mechanical properties, e.g. hardening and embrittlement. The irradiation program ARBOR was performed at the BOR60 reactor of State Scientific Center - Research Institute of Atomic Reactors (SSC-RIAR) in Dimitrovgrad (Russia) [37], investigating mainly dose dependence of radiation effects. Tab. 2.1 shows important specifications of irradiation experiments SPICE and ARBOR, on which simulation efforts in this work have been based. In SPICE, one irradiation cycle was performed to a cumulative maximum damage of 16.3 dpa, with specimens positioned at different levels of the reactor core, thus leading to different irradiation temperatures in steps of 50 °C and varying doses. A two-step irradiation cycle in ARBOR1 and ARBOR2 led to doses of around 32 and 70 dpa, respectively, while

Table 2.1.: Irradiation programs with main specifications. Thermal neutrons correspond to $E < 0.68\,\mathrm{eV}$, fast neutrons to $E > 0.1\,\mathrm{MeV}$.

	SPICE	ARBOR 1 & 2
Irradiation facility	HFR, Petten	BOR60, Dimitrovgrad
Neutron flux ($\mathrm{m^{-2}s^{-1}}$)	1.4×10^{18} (thermal)	–
	4.0×10^{18} (fast)	1.8×10^{19} (fast)
Irradiation time (10^7s)	6.67	3.98 & 8.68
Cumulative dose (dpa)	16.3	32 & 70
Irr. temperature ($^\circ$C)	250 - 450	330 - 340

Table 2.2.: Displacement damage and gas production rates in RAFM steels under neutron irradiation from different sources [39].

Irradiation parameter	Fusion FW (3-4 GW reactor)	IFMIF	HFR	BOR60
Damage rate (dpa/year)	20-30	20-55	$\sim$7	$\sim$20
H (appm/dpa)	40-50	40-50		$\leq$10
He (appm/dpa)	10-15	10-12	$\leq$1	$\leq$1

other parameters remained constant. HFR and BOR60 differ significantly in their neutron spectra: compared to ARBOR, SPICE specimens suffered from irradiation of fast as well as thermal neutrons, which altered occurrence and strength of radiation effects. Tab. 2.2 summarizes damage rates and gas production rates for several irradiation facilities, which are compared to expected fusion conditions. While the damage rate of BOR60 fulfills fusion requirements, gas production is too low. The neutron spectrum of HFR is also not capable of copying the irradiation effects of fusion. Answering this purpose, the International Fusion Materials Irradiation Facility (IFMIF) is planned to be built, which, by using two 40 MeV deuteron accelerators and a flowing lithium target, will be able to produce a fusion like neutron spectrum [38] for further irradiation studies. Damage rate and gas production of IFMIF are also shown in Tab. 2.2, coping with fusion expectations.

Several specimen types fabricated out of different materials were irradiated in experiments SPICE and ARBOR. Samples were mechanically tested before and after irradiation by impact, tensile and Low Cycle Fatigue (LCF) tests. Concerning the material, besides conventional RAFM steels, model alloys doped with boron were exposed to neutron radiation, whereby helium production was artificially increased due to a boron-helium transmutation in order to simulate fusion gas production more realistically. This experimental simulation method is explained in Section 2.5.1, while the specifications of boron doped steels are described in Section 3.4.

2.5. Experimental helium effects simulation methods

Conventional fission experimental reactors, as mentioned earlier, show, due to their distinct neutron spectra, different characteristics of radiation effects and in particular helium generation. Several experimental simulation techniques exist to enhance helium production in RAFM steels up to levels that are expected for a FW of a fusion reactor. These methods are presented here.

2.5.1. Transmutation of dopants

Additional alloying elements, which are not contained in the basic steel composition, are used to create helium by interacting with impinging neutrons. Candidates for this technique are elements nickel and boron. Under neutron irradiation, the following processes take place:

$$^{58}\text{Ni (n,}\gamma\text{) }^{59}\text{Ni (n,}\alpha\text{) }^{56}\text{Fe,} \tag{2.2}$$

where natural Ni contains an atomic fraction of 68% ^{58}Ni, which undergoes first a gamma and then an alpha decay; and

$$^{10}\text{B (n,}\alpha\text{) }^{7}\text{Li,} \tag{2.3}$$

where ^{10}B (natural abundance of 20%, 80% ^{11}B) directly produces helium and lithium.

Advantages of these techniques are that these dopants are highly receptive even for thermal neutron irradiation that is produced by experimental fission reactors. Low activation energies for ^{10}B and ^{58}Ni transmutation yield high helium generation rates while H production, in principle, is lower than expected under fusion conditions. For F/M steels, a nickel content of about 2% Ni would increase helium production up to 10 appm/dpa [40] in a mixed-spectrum reactor like HFR, which fits quite well the fusion expectations in Tab. 2.2. Unfortunately, due to Ni activation, structural RAFM steels are especially designed without Ni. Re-alloying Ni, even for the purpose of artificial helium generation, would again introduce Ni effects on the microstructure and mechanical properties, which should be avoided.

Boron, on the other hand, does not produce radiant isotopes by transmutation. A drawback of this technique is a not yet identified role of the reaction product lithium. Nevertheless, lithium effects were shown to be minor by comparing the DBTT shifts in neutron irradiated boron doped and helium implanted F82H steel specimens in [33]. ^{10}B has a large reaction cross section for thermal neutrons (e.g. in HFR), while fast neutrons (e.g. in BOR60) cause a low transmutation rate. By adjusting the ^{10}B fraction, it is possible to cast model alloys with the same concentration of boron, but different final helium concentrations and generation rates, minimizing the influence of the microstructure on the mechanical properties. Important for doping techniques, in general, is the homogeneous distribution of the dopant in the steel, which ensures that also helium is generated homogeneously in the matrix. Unfortunately, the fate of boron was unclear, whether e.g. boron precipitates are formed or boron segregates at grain boundaries. Since boron doping is the method that had been applied to RAFM steel EUROFER97 specimens investigated in this work, their microstructure was analyzed by Auger Electron Spectrometry (AES) and TEM concerning boron distribution in the unirradiated state (see Section 3.4.3.1).

2.5.2. Helium implantation

Helium ion (α particle) irradiation causes implantation of helium into the steel matrix. α particle sources are e.g. the *Jülich cyclotron*, where ^{4}He ions are accelerated to energies up to 28 MeV [41]. Unfortunately, specimen thickness is limited to a maximum of 100 μm to achieve homogeneous implantation through the whole specimen volume. α particle irradiation causes high helium-to-dpa ratios and extremely high helium production rates, which exceed fusion expectations by more than two orders of magnitude (see also [42]). By combination of α particle with heavy ion irradiation through dual beam injection or subsequent irradiation, helium generation and damage dose can be adjusted to levels which are more fusion alike.

2.5.3. Surface implantation coating

Another method of introducing helium into a specimen is by using a surface coating, e.g. with nickel [43]. Neutron irradiation by experimental fission reactors causes nickel-helium transmutation according to Eq. (2.2), and the generated helium is injected into the sample by impact processes due to the neutron bombardment. The helium-to-dpa ratio is adjustable, but the depth of uniform helium implantation is limited to a few micrometers, which disqualifies this method for bulk samples.

2.5.4. Spallation neutron sources

In spallation neutron sources, e.g. SINQ of Paul Scherrer Institute in Switzerland [44, 32], high energy particles like protons are accelerated at a target, e.g. lead. At the impact, neutrons with high kinetic energy are ejected which are used to irradiate specimens. For the SINQ facility, the proton beam is accelerated up to 630 MeV, and is directed towards the lead target. Resulting neutrons can be moderated to achieve the desired neutron energy for further experiments. Compared to fusion, spallation neutron sources show much higher helium-to-dpa ratios [40, 42]. Furthermore, high concentrations of hydrogen are created causing severe embrittlement of the irradiated material [45].

2.6. Modeling and simulation techniques

Facing new developments, modeling and simulation techniques nowadays present an important tool. They are capable of describing macroscopic material properties, e.g. tensile behavior, as well as providing the source for this down to the atomic level, e.g. through atom-atom interactions. Besides different length scales, effects are aimed to be described from picoseconds up to years. However, proper simulation techniques need to be chosen to bridge length and time scales. In

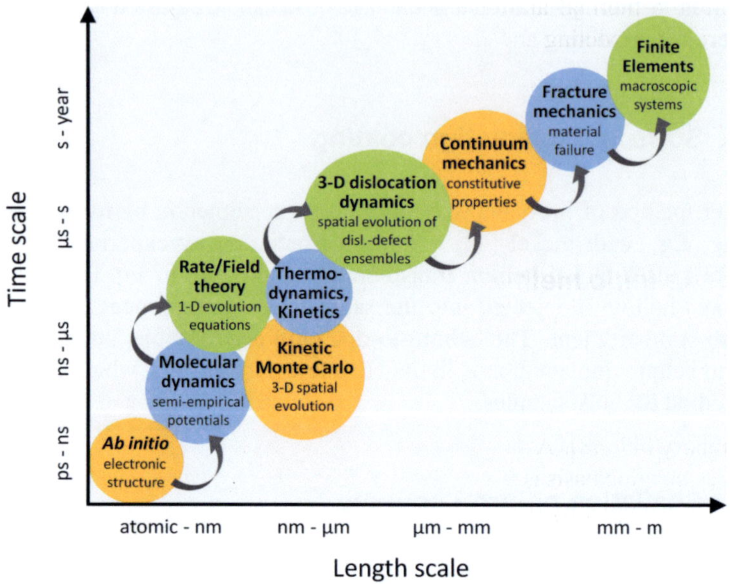

Figure 2.8.: Modeling and simulation methods classified into their associated time and length scale [46]. Together they present a possible multiscale modeling approach, where results of one method serve as input parameters for successive techniques.

Fig. 2.8 several modeling and simulation approaches are shown, classified into the appropriate length and time scale [46].

Obviously, one simulation technique can not be valid for the whole range, because models for small time and length scales would consume large computer resources, yet result in high calculation times. Therefore, a chain of simulation methods, in which each rely on the results of calculations from other techniques, has to be built, thus achieving a multiscale modeling approach.

Concerning fusion, modeling and simulation gain even more importance since irradiation experiments are not able to reproduce expected neutron effects from fusion conditions. This is because for the time being no irradiation facility exists which is capable of producing a fusion–like neutron spectrum. Therefore, neutron effects in fusion have to be estimated and extrapolated from measured irradiation influences in thermal and fast neutron sources (or other particle irradiation), which is why modeling and simulation methods are necessary.

In the following paragraphs, modeling techniques and simulation methods are explained whose results served as preliminary investigations and input data for the developed model. Detailed results necessary for this work are presented in Section 3.2.

2.6.1. *Ab initio* methods

A quantum mechanical approach provides the basis for *ab initio* methods. Therein, atom-atom interactions are described by solving their electron wave functions (Schrödinger equations), using certain approximations like e.g. the Hartree-Fock theory [47, 48]. A second approach is Density Functional Theory (DFT), whose calculation basis is the electron density, and not the separate electron wave functions. Complexity of calculations is less, due to less degrees of freedom of the electron density. While the basic theory of Thomas, Fermi and Dirac [49, 50] limits DFT to certain special problems, modifications such as Hohenberg-Kohn theorems [51] and Kohn-Sham theory [52] partly solved this problem. DFT allows treatment of many-electron systems by using appropriate electron density approximations and pseudopotentials. Results of both calculations lead to the energy ground state of the (molecular) system, whereof structural and energetical information, e.g. coordination and binding energies, is accessible.

2.6.2. Molecular Dynamics simulation

Molecular Dynamics (MD) simulations [53] are capable of describing effects on atomic (or molecular) scale in the solid, liquid or gaseous state for a typical time range of pico- or nanoseconds. A simulation box is defined, which represents the volume or area of interest. The interactions (forces) between the enclosed atoms are now described by semi-empirical pair potentials, which reduce the many-body problem to the sum over all interacting pairs. Repelling or attracting forces between each atom within a certain range characteristic for each atom-atom interaction are calculated. Forces yield a net movement in a specific direction for every atom resulting in an equlibrated state after a certain simulation time. Taking this state as initial situation, further simulations can now be performed.

As an example, displacement cascades due to neutron impingement are simulated by giving a randomly chosen steel lattice atom a certain velocity in one direction; and the effects on the other atoms in the calculation box are studied [54].

Speaking of a two dimensional area, e.g. a plane of lattice atoms, the simulation container has to be large enough, so that effects of the walls or boundaries are able to be neglected. To achieve this, typically periodic boundary conditions are applied. This is, that if an atom moves out of the container, lets say through the right wall, it has to be introduced through the left wall at the same moment, having the same velocity and direction. Due to additional other long range interactions depending on the purpose of the simulation, container limits have to be chosen with care, and simulation results have to be checked for unintended boundary effects.

2.6.3. Monte Carlo simulation

Monte Carlo (MC) simulations [55, 56] are applicable to systems, in which the evolution of processes are based on stochastic probabilities. Comparable to MD, a simulation box is defined, building the basis for MC simulations. Calculations are performed on the nanometer scale, typically using periodic boundary conditions for the simulation box. Starting from a chosen initial position of the atoms, the energy of the whole system is calculated from interacting potentials and memorized. Subsequently, one atom is moved by a small distance, and the energy of the system is recalculated. If the system has lowered its energy, the new configuration is accepted and provides a new source for a next calculation step. If the system's energy increases, a random number is generated which decides if the configuration is nonetheless accepted. By this means, energetically stable system configurations will be found, even if the starting point was close to a metastable local energy minimum.

Besides MC simulations, which describe quasi steady state processes leading to an equilibrated state, kinetic Monte Carlo (kMC) simulations are capable of describing time dependent dynamic events, which is important for diffusion processes. In addition to MC, kMC simulations need a correlation of the simulation time to the real time of the experiment. A method called *first passage time analysis* [55] is used, which allows calculation of system changes for small real time steps Δt.

2.6.4. Rate/Field Theory

Considering a system of different kinds of microstructural species, their interaction and reactions can be described by a Rate or Field Theory model. The changes of the initial concentration of each kind of particle are described by rate equations, which increase or decrease the actual concentration due to generation, mutual reactions or loss of particles. Depending on the system size and number of different species, a large amount of rate equations appears. Since this system of equations is rarely possible to be solved analytically, a numerical approach has to be used.

In the following a brief description of this method is presented as an example. In [57] rate equations are used to describe the evolution of SIA loops growing during irradiation. SIAs are generated together with vacancies during irradiation, forming at first small clusters, and later on dislocation loops and voids, respectively, by undergoing a mutual reaction and growth process. A rate equation has to be set up for every defect type and size, where all possible interactions are described. Under the assumption that only single defects are mobile the concentration of single SIAs in the matrix is taken as an example (see [57, Eq. (4)]), which is increased by

- the defect production due to the irradiation,

- the reaction of a vacancy with a SIA cluster of size 2.

The concentration of single SIAs is decreased by

- the annihilation due to reactions of a vacancy with a SIA,

- reactions of single SIAs,

- every attachment of a single SIA to a SIA cluster or loop,

- every reaction of a single SIA with a vacancy cluster,

- the amount of SIAs lost to microstructural sinks e.g. line dislocations.

Every reaction happens with a certain probability expressed by kinetic rates, which are derived from thermodynamic and/or geometric considerations. In case of the model in [57], equations for clusters of sizes larger than 4 were translated to Fokker-Planck Equations (FPEs), which are partial differential equations consisting of a diffusion and drift term [58]. These two effects lead to a broadening and shift of the resulting cluster size distribution with time.

Obviously, the number and complexity of the rate equations to be solved is high, demanding sophisticated programming and numerical solving methods to optimize the calculation. The model presented in this work is also based on Rate Theory, and will be explained in detail in Chapter 4.

3. State-of-the-art of research

3.1. Thermodynamic considerations

3.1.1. Nucleation theory

The model developed in this work is based on the theory of homogeneous nucleation [59, 60]. A cluster represents an enclosed consistent volume, which is separated from the surrounding matrix due to different characteristics like composition, building a second phase and interface. The work necessary to form a cluster of a second phase depends on the cluster size. Fig. 3.1 exemplarily shows the formation energy expressed through changes in the Gibbs free energy G of the cluster of radius R analog to

$$\Delta G = -G_{\mathrm{vol}} R^3 + G_{\mathrm{surf}} R^2, \tag{3.1}$$

with G_{vol} and G_{surf} describing a volume and surface parameter, respectively.

Two opposing effects are observed: while the creation of a new uniform volume decreases the Gibbs free energy, the formation of the interface between the two phases leads to an increase of G. Since the surface energy term G_{surf} scales with R^2 and the volume energy term G_{vol} with R^3, clusters need a certain critical size R_{cr}: for $R < R_{\mathrm{cr}}$ clusters are energetically more favored to shrink, whereas clusters larger than R_{cr} experience stable growth. Volume and surface energy terms depend on environmental conditions such as temperature and supersaturation of the second phase forming species, resulting in the decay of a supersaturated solid solution towards equilibrium concentration.

As solubility of helium in the steel matrix is very low, a high helium supersaturation is achieved due to helium transmutation by neutron irradiation. For this reason, clusters are assumed to be stable starting from a cluster size of two helium atoms, and furthermore undergo stable growth. This so called diatomic nucleation

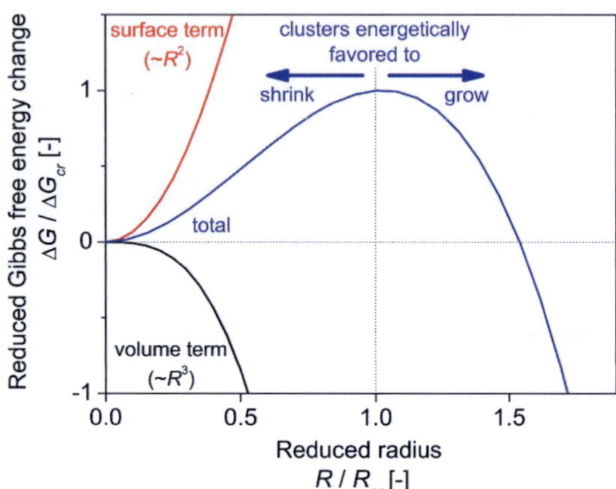

Figure 3.1.: Gibbs free energy change during nucleation of a spherical second phase with radius R analog Eq. (3.1). At a critical radius R_{cr}, Gibbs free energy G passes through a maximum, larger clusters are energetically favored to grow stable.

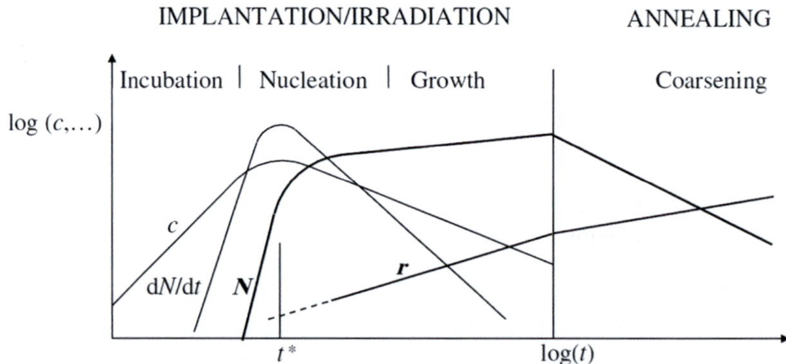

Figure 3.2.: Schematically depicted time dependence of main quantities characteristic for helium bubble nucleation and growth presented by Trinkaus [61]. The notation slightly differs from the variables used in this work: Evolution of solved helium concentration c, cluster density N, nucleation rate $\mathrm{d}N/\mathrm{d}t$ and mean cluster radius r is shown for continuous helium generation under irradiation. Due to the development of the curves, different stages are distinguished. Finally, curve behavior under the annealing condition is shown.

mechanism is widely accepted for helium bubbles and well described in literature, e.g. in [61].

Fig. 3.2 is derived from Trinkaus [61], showing a basic scheme of the typical evolution of the main quantities characteristic for helium bubble nucleation and growth. Under irradiation and continuous helium generation, three time domains can be distinguished:

1. In the incubation stage, the concentration of solved helium, c, builds up due to helium generation by transmutation.

2. In the nucleation stage, supersaturation of helium is high enough to start the di-atomic nucleation mechanism. The nucleation rate, $\mathrm{d}N/\mathrm{d}t$ passes through a maximum, which coincides with the maximum of c. Cluster density, N steeply increases, after the maximum of the nucleation rate, newly generated helium is more and more likely to be added to existing clusters than to find another single solvent helium atom to nucleate a new bubble. Mean cluster radius, r increases with further helium generation.

3. In the growth stage, under constant helium generation, N and r steadily increase, while c and $\mathrm{d}N/\mathrm{d}t$ decrease.

After irradiation, the effects of an annealing period is shown: helium generation has stopped, which leads to coarsening by e.g. coalescence or Ostwald ripening of bubbles, reducing the cluster density N and increasing the mean cluster radius r.

The model, which is developed in this work, is to follow the mentioned specifications under irradiation. Simulation results will be compared to the scheme in Fig. 3.2.

3.1.2. Cavity growth

By definition, a cavity can either be gas-filled, e.g. with helium, and is called a bubble, or only be a cluster of vacancies which is called a void. Void growth has been the subject of many scientific studies due to the severe problem of swelling in austenitic stainless steels under neutron irradiation in fission reactors. Investigations showed that in the presence of helium small voids are stabilized and the critical cluster radius for a stable nucleus decreased.

Generally, the growth of a cavity with volume V^V can be described by the fluxes of the void forming point defects, vacancies and Self Interstitial Atoms (SIAs), directed from or towards the void as shown in [62], with their corresponding atomic volume Ω analog to

$$\frac{\mathrm{d}V^V}{\mathrm{d}t} = \frac{\mathrm{d}}{\mathrm{d}t}\left(\frac{4}{3}\pi R^3\right) = \Omega\left[4\pi R D_\mathrm{v}\left(C_\mathrm{v}-C_\mathrm{v}^\mathrm{V}\right) - 4\pi R D_\mathrm{i}\left(C_\mathrm{i}-C_\mathrm{i}^\mathrm{V}\right)\right], \qquad (3.2)$$

and finally the radial growth rate of the void with radius R is given by

$$\frac{\mathrm{d}R}{\mathrm{d}t} = \frac{\Omega}{R}\left[D_\mathrm{v}\left(C_\mathrm{v}-C_\mathrm{v}^\mathrm{V}\right) - D_\mathrm{i}\left(C_\mathrm{i}-C_\mathrm{i}^\mathrm{V}\right)\right]. \qquad (3.3)$$

C_v^V and C_i^V are vacancy and interstitial concentrations at the void surface, and C_v and C_i in the bulk volume, respectively, and D_v and D_i are the diffusion coefficients of vacancies and SIAs. When introducing gas into the void, a bubble is formed with an inner pressure p. Point defect concentrations on the bubble surface can be calculated by using thermodynamic equilibrium concentration of vacancies, C_v^0:

$$C_\mathrm{v}^\mathrm{V} = C_\mathrm{v}^0 \exp\left[-\frac{\Omega}{k_\mathrm{B}T}\left(p-\frac{2\gamma}{R}\right)\right], \qquad (3.4)$$

and corresponding interstitial equilibrium concentration, C_i^0:

$$C_\mathrm{i}^\mathrm{V} = C_\mathrm{i}^0 \exp\left[\frac{\Omega}{k_\mathrm{B}T}\left(p-\frac{2\gamma}{R}\right)\right]. \qquad (3.5)$$

Therein, γ is the surface energy of the matrix material, T is the temperature and k_B is the Boltzmann constant. That is, for a bubble in equilibrium, i.e. with pressure $p = 2\gamma/R$, the vacancy concentration at the bubble surface equals the bulk equilibrium concentration. The interstitial equilibrium concentration is very low, which is why C_i^V in Eq. (3.3) can be neglected [62]. Setting $\mathrm{d}R/\mathrm{d}t = 0$ and inserting Eq. (3.4), the critical radius can be calculated as

$$R_\mathrm{cr} = \frac{2\gamma}{p+\frac{k_\mathrm{B}T}{\Omega}\ln S_\mathrm{v}}, \qquad (3.6)$$

with S_v being the effective vacancy supersaturation given by

$$S_\mathrm{v} = \frac{D_\mathrm{v}C_\mathrm{v}-D_\mathrm{i}C_\mathrm{i}}{D_\mathrm{v}C_\mathrm{v}^0}. \qquad (3.7)$$

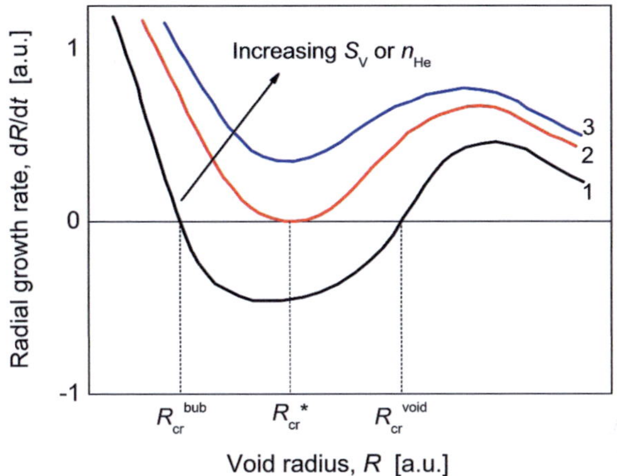

Figure 3.3.: Radial void growth rate depending on void radius [63, 62]. An increasing vacancy supersaturation S_v or amount of helium atoms in the void n_He supports a bias-driven growth by vacancy absorption, where no further helium is needed to stabilize the void.

The bubble pressure p can be expressed by an equation of state, and in the case of an ideal gas approximation it can be correlated to the radius of the spherical bubble and included gas atoms by the ideal gas law through

$$p = \frac{3 n_\mathrm{He} k_\mathrm{B} T}{4\pi R^3}.$$
(3.8)

By inserting Eq. (3.8) in (3.6), a function $g\left(R_\mathrm{cr}\right)$ can be derived:

$$g\left(R_\mathrm{cr}\right) = R_\mathrm{cr}^3 - \frac{2\gamma\Omega}{k_\mathrm{B} T \ln S_\mathrm{v}} R_\mathrm{cr}^2 + \frac{3 n_\mathrm{He}\Omega}{4\pi \ln S_\mathrm{v}}.$$
(3.9)

Fig. 3.3 shows the typical evolution of the radial growth rate with cluster size for different parameter values [62, 63]. Curve 1 has two roots in the relevant regime: bubbles smaller than $R_\mathrm{cr}^\mathrm{bub}$ grow until they reach the critical size. If a further vacancy is attached, it is highly probable that this vacancy is emitted again because the growth rate becomes negative. If, by statistical perturbations, the bubble reaches $R_\mathrm{cr}^\mathrm{void}$ the bubble experiences bias-driven growth by vacancy absorption. The higher the amount of helium atoms in the bubble is, the smaller is the difference between the two roots. From a critical cluster radius $R_{cr}^{\star}$ on, no negative growth rate exists any more (curve 2), meaning that from this point the bubble suffers unstable growth without need for attaching further helium atoms. This case is demonstrated by the curve 3 where the growth rate is positive for all cluster sizes.

The critical number of helium atoms in the bubble, $n_{He}^{\star}$ which allows the bubble to grow unstable by vacancy absorption, can be calculated by building the derivative with R_{cr} of Eq. (3.9), and setting the derivative together with Eq. (3.9) to zero: calculations therefore find the root of $g(R_{cr})$ where R_{cr} is also a local minimum analog to curve 2 in Fig. 3.3, yielding

$$n_{He}^{\star} = \frac{32 F_v \gamma^3 \Omega^2}{27 (k_B T)^3 (\ln S_v)^2},$$

(3.10)

with F_v being a shape factor which becomes $4\pi/3$ for spherical bubbles. The corresponding critical radius can be derived as

$$R_{cr}^{\star} = \frac{4\gamma\Omega}{3 k_B T \ln S_v}.$$

(3.11)

A higher vacancy supersaturation supports the change to unstable void growth by decreasing $n_{He}^{\star}$ and $R_{cr}^{\star}$ (illustrated by the arrow in Fig. 3.3).

As a consequence of unstable bias driven void growth, cavity size distributions show a bimodal size distribution. While bubble sizes peak at smaller diameters, a few cavities, which exceed the critical cluster radius, form a second distribution with a larger peak diameter. These bimodal cavity size distributions have been reported especially for the intensely investigated austenitic stainless steels (see summarizing table in [64]).

3.2. Review of simulation results

3.2.1. Helium diffusivity

For the simulation of helium bubble nucleation and growth it is of great importance to understand and evaluate helium diffusion mechanisms taking place

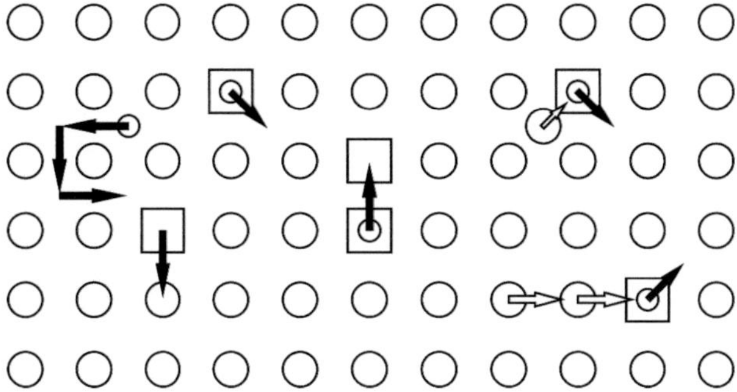

Figure 3.4.: Scheme of possible diffusion mechanisms relevant for helium migration [61]. Large circles symbolize iron atoms, small circles helium atoms, squares represent vacancies.

especially under irradiation conditions. Generally, Fig. 3.4 shows possible diffusion mechanisms [61] relevant for helium migration, which from left to right are:

- helium interstitial migration,

- vacancy migration,

- dissociation of a substitutional helium atom by a thermally activated jump to an interstitial position,

- substitutional helium jumps to a nearest neighbor vacancy (helium-divacancy mechanism),

- iron SIA pushes a substitutional helium atom to an interstitial position,

- cascade induced helium dissociation.

All mechanisms may be more or less involved under neutron irradiation, which is why an effective helium diffusivity has to be defined and estimated.

Calculations of helium energetics in body-centered cubic (bcc) iron are used to get an idea of the most likely diffusion mechanism. The migration energy of interstitial helium He_1 is quite low (0.06 eV) [65]. Nevertheless, helium has a high tendency to be trapped by vacancies (V), which leads to substitutional helium ($He_1 V_1$) with a high binding energy of 2.3 eV. Possible diffusion paths are

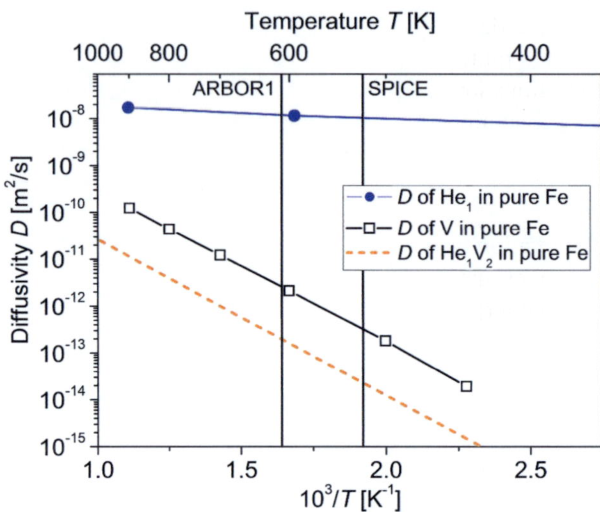

Figure 3.5.: Temperature dependent diffusivities of interstitial helium He$_1$, vacancies V [67] and helium-divacancies He$_1$V$_2$ [68] under neutron irradiation condition. Vertical lines show relevant temperatures of SPICE and ARBOR1 irradiation experiments.

then given by a dissociation mechanism, i.e. jumping to an interstitial position and further interstitial migration, or by migration via vacancies. Migration energies thereby depend strongly on the vacancy concentration. For a high vacancy supersaturation, which is expected due to high point defect generation at low temperature irradiation conditions, migration through a helium-divacancy mechanism owns the lowest energy of 0.3 eV [66], which is therefore assumed to be the main diffusion mechanism taking place. Nevertheless, other vacancy mechanisms may participate in helium diffusion leading to a vacancy-mediated helium diffusion mechanism, which is thus described in the model. At high temperatures, on the other hand, when thermal vacancies dominate, the dissociation mechanism has the lowest effective migration energy of 0.24 eV, while for a helium-divacancy mechanism the migration energy was calculated to be 2.36 eV.

Fig. 3.5 shows temperature dependent diffusion coefficients of interstitial helium, vacancies and helium-divacancy clusters. Calculated by atomistic Molecular Dynamics (MD) simulations [68] and kinetic Monte Carlo (kMC) simulations using *ab initio* input data [67] the curves show the same trend: while interstitial helium diffusion is quite fast, trapping of helium by vacancies under neutron irradiation conditions lowers its diffusivity below the diffusion coefficient for single vacancies assuming a helium-divacancy mechanism.

Irradiation leads to an increase in the effective helium diffusivity. The defect concentrations, especially vacancy concentration, are highly enhanced due to Frenkel pair generation by displacement cascades. This immediately leads to a supersaturation of vacancies largely above thermal equilibrium, because SIAs as corresponding counterparts are more mobile and easily form dislocation loops or are trapped by sinks. Therefore, vacancy supersaturation increases helium diffusivity by providing more effective vacancy mediated diffusion paths. Besides, even new paths can be produced by defect species, which are not present or stable under equilibrium conditions, also being responsible for Radiation Enhanced Diffusion (RED) [62, 69].

3.2.2. Helium density in bubbles

Morishita [70] used MD and Molecular Static (MS) simulation techniques to investigate the state of helium bubbles in bcc iron for various contents of helium atoms, SIAs and vacancies. Assuming that the formation free energy of helium bubbles is independent of temperature, calculations were performed to determine helium bubble formation energies up to helium and vacancy numbers of 100. Fig. 3.6a shows evolution of helium bubble formation energies for a fixed helium content with increasing number of vacancies. Results yield lowest formation energies for helium bubbles with a helium-to-vacancy ratio of approximately one. Fig. 3.6b presents binding energies of point defects to helium bubbles with different helium-to-vacancy ratios, for bubble sizes of less than 20 helium atoms and vacancies. For ratios less than one, point defect binding energies are almost constant. For ratios from one to six, vacancy binding energy increases, and SIA and interstitial helium binding energy decreases. For ratios larger than six, binding energies show reversed tendencies, which was ascribed to relaxation effects of the iron lattice. It was shown that point defect binding energies depend on the helium density in the bubble, and not the bubble size. It can be observed, that for a helium-to-vacancy ratio between one and two point defect binding energies coincide, leading to the assumption that those bubbles are most stable.

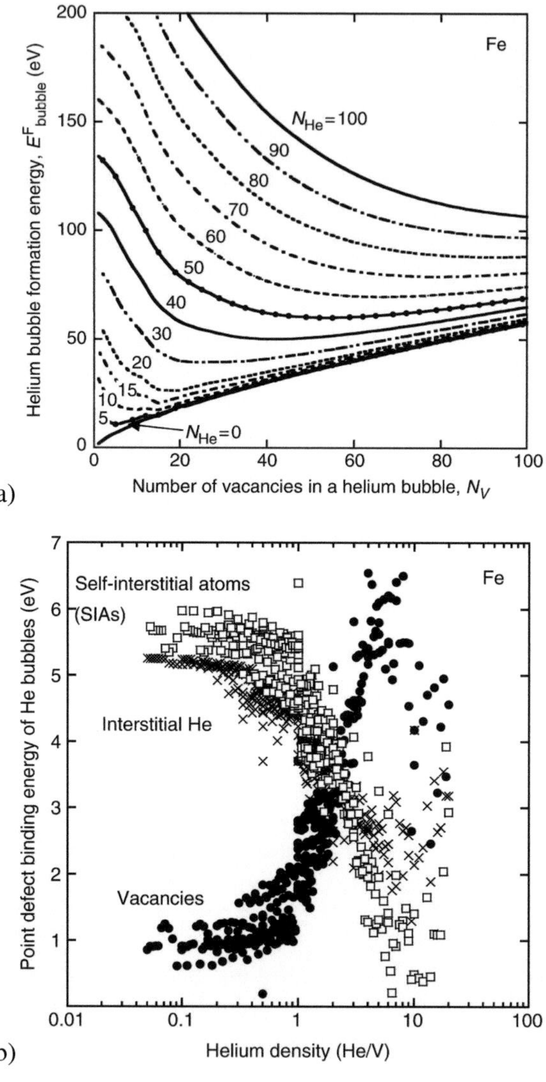

Figure 3.6.: MD simulations by Morishita [70]. a) Lowest helium bubble formation energies are observed for helium-to-vacancy ratios of approximately unity. b) Point defect binding energies to bubbles depend on helium-to-vacancy ratio, with similar binding energies for SIAs and vacancies for ratios between one and two.

Hafez Haghighat *et al.* [71] performed MD simulation on a pressurized helium bubble with diameter of 2 nm in bcc iron for different temperatures. The helium-to-vacancy ratio inside the bubble was varied. For a ratio of one, helium atoms leave a small gap between themselves and the iron lattice atoms due to repulsing forces; the bubble pressure was calculated to 8 GPa at 300 °C. An increase of the helium-to-vacancy ratio from 1 to 2 yield a steep helium pressure increase to 27 GPa. At higher ratios the iron lattice even starts to deform plastically. It can be stated that helium-to-vacancy ratios of two and above are unlikely to occur due to the associated large bubble pressures for this bubble size.

3.3. Modeling of hardening

The mechanical strength of a material is defined as the resistance towards plastic deformation. The main mechanism of plastic deformation is based on dislocation glide, which needs a characteristic shear stress to start moving through the lattice. Lattice defects in the slip plane of the moving dislocation serve as obstacles, which either have to be cut or bypassed. In both cases, a higher shear stress is necessary to provide enough energy for the dislocation to further glide, until then this dislocation is pinned and does not contribute to plastic deformation.

Under irradiation, lattice defects like dislocation loops, voids or bubbles are generated, which additionally impede dislocation glide and therefore lead to irradiation hardening and thus to an increase in yield strength. A correlation of microstructural defects with hardening can be done by applying an appropriate hardening model, e.g. the Dispersed Barrier Hardening (DBH) model [72, 73, 74], to measured or simulated defect sizes and densities. Within this model, obstacles are considered homogeneously distributed in the matrix. Their mutual spacing, l_j is the determining factor increasing the yield strength. In principle, hardening is proportional to l_j^{-1}, and can be calculated for a defect type j by

$$\Delta\sigma_j = M_{\mathrm{T}}\alpha_j\mu b_{\mathrm{disl}}l_j^{-1} = \begin{cases} M_{\mathrm{T}}\alpha_j\mu b_{\mathrm{disl}}\sqrt{N_j d_j} & \text{for discrete obstacles } j, \\ M_{\mathrm{T}}\alpha_{\mathrm{LD}}\mu b_{\mathrm{disl}}\sqrt{\rho_{\mathrm{LD}}} & \text{for line dislocations.} \end{cases}$$

$$(3.12)$$

Further parameters are shear modulus of the matrix μ, Burgers vector b_{disl} of gliding dislocations, barrier strength α_j, obstacle density N_j and size d_j, dislocation density ρ_{LD}. The Taylor factor M_{T} relates the shear stress necessary for a dislocation to glide on a slip plane in a single crystal to the external tensile stress applied to a polycrystal. M_T equals 3.06 for the bcc lattice [75]. The barrier strength α_j varies from 0.15 for weak obstacles like line dislocations, 0.6 for intermediate obstacles like dislocation loops, up to 1 for strong obstacles like voids [72]. The barrier strength of helium bubbles is assumed to lie in the weak/intermediate range from 0.2 to 0.4.

Following the mentioned principle, hardening of one-sized particles, e.g. oxide particles in oxide dispersion strengthened (ODS) steels, with diameter d and density N according to Eq. (3.12) can be assessed. Accounting for hardening effects of different obstacle types j, a superposition law [72] has been proposed:

$$\Delta\sigma_{tot} = \sqrt{\sum_j \left(\Delta\sigma_{SR,j}\right)^2} + \Delta\sigma_{LR}, \qquad (3.13)$$

where obstacles are divided into short range (SR) and long range (LR) obstacles due to their effects on lattice distortion. SR obstacles directly influence dislocation movement only on the dislocation glide planes, while LR obstacles produce long range induced stresses on elastic lattice properties. While hardening from SR obstacles, like voids, bubbles and dislocation loops, is summed up and added to the total hardening $\Delta\sigma_{tot}$ by using a root sum square (RSS) rule, hardening due to LR obstacles, like line dislocation networks, is added by linear superposition.

3.4. RAFM steels and boron doped model alloys at KIT

3.4.1. Composition and irradiation characteristics

For a good economical and environmental use fusion technology demands new structural materials, which are capable of surviving a long time under the harsh fusion conditions, without need for replacement due to deteriorating properties, and without development of high activation. Therefore, and owing to the sensitivity to neutron irradiation of austenitic stainless steels (see Section 2.2), the new class of Reduced Activation Ferritic/Martensitic (RAFM) steels with bcc lattice structure had been developed and has been studied so far. Main purpose was to achieve comparable mechanical properties, better radiation resistance and lower activation than for austenitic stainless steels. Because nickel, one of the main alloying elements of stainless steels, transmutes under neutron irradiation to long living radiant isotopes, it had to be replaced together with elements niobium and molybdenum. Elements tungsten, vanadium and tantalum were added to the steel composition, which, concerning mechanical properties, undertake the tasks of the omitted elements: mainly increasing strength at low and elevated temperatures by building carbides. EUROFER97, developed mainly by *Institute for Materials Research* at Forschungszentrum Karlsruhe (FZK) [76, 77, 78] (now Karlsruhe Institute of Technology (KIT)) based on research from European institutions, is the European type of RAFM steels, with composition given in Tab. 3.1. Variations in shown compositions exist because 10 heats with a total mass of 11 tons were cast to different product forms (forged bars, plates, tubes, wires). Several undesired elements were also included due to the casting process.

In addition, boron doped model alloys, named ADS2 to ADS4, based on the EUROFER97 composition were cast [77]. Variations of boron isotope ^{10}B, which is highly receptive to transmuting to helium under neutron irradiation from conventional fission experimental reactors, were used to simulate helium production closer to fusion expectations. As described in Fig. 2.4c, by an (n,α)-reaction ^{10}B transmutes to ^{7}Li and ^{4}He. While ADS2, like EUROFER97, contains natural boron, whereof ^{10}B takes a fraction of 20%, ADS3 and ADS4 were doped with different concentrations of separated ^{10}B isotope (see Tab. 3.1). Noteworthy, ADS2 and ADS3 contain the same amount of the element boron (but not the same amount of ^{10}B), which makes it possible to directly assess the effect of the transmuted helium concentration, without having differences in steel composition and microstructure before irradiation.

After irradiation, however, the microstructure had changed. Tab. 3.1 shows cumulative helium concentrations achieved in SPICE and both ARBOR experiments. Damage doses differ from specifications in Tab. 2.1 because data corresponds to impact specimens, which due to their position in the reactor did not suffer the highest doses in SPICE and ARBOR1. One can observe that the different neutron spectra have a strong influence on the helium production. For SPICE performed in High Flux Reactor (HFR), the highest helium concentrations are achieved, although the damage dose is the lowest. As all boron was burnt up already within 2 dpa, the helium generation rate was not constant over the whole time of the SPICE irradiation experiment. This makes it difficult to compare SPICE with fusion conditions although the final helium amount achieved in ADS3 fits expectations. In contrast, fast neutrons in the ARBOR experiment cause quite

Table 3.1.: Composition of RAFM steel EUROFER97 and boron doped model alloys ADS2 to ADS4 [76, 77, 23] (Fe balance). Values show range over 10 different heats, with a total weight of 11 tons. Product forms include forged bars, plates, tubes and wires.
In the lower part, final helium concentrations in EUROFER97 and boron doped alloys due to boron transmutation in SPICE and ARBOR are shown.

Main alloying elements	EUROFER97 (mass%)	ADS2 (mass%)	ADS3 (mass%)	ADS4 (mass%)
C	0.11 - 0.12	0.109	0.095	0.100
Cr	8.82 - 8.96	9.31	8.80	9.0
W	1.07 - 1.15	1.27	1.125	1.06
Mn	0.38 - 0.49	0.602	0.395	0.38
V	0.18 - 0.20	0.19	0.193	0.197
Ta	0.13 - 0.15	0.055	0.088	0.08
N	0.018 - 0.034	0.021	0.028	0.0255
P	0.004 - 0.005	0.0035	0.0024	0.001
S	0.003 - 0.004	0.0030	0.003	0.0025
natural B	0.0005 - 0.0009	0.0082		
^{10}B			0.0083	0.112
O	0.0013 - 0.0018	0.013	0.0045	0.0037
Radiologically undesired elements	(μg/g)			
Nb	2 - 7			
Mo	10 - 32			
Ni	70 - 280			
Cu	15 - 220			
Al	60 - 90			
Ti	50 - 90			
Si	400 - 700			
Co	30 - 70			
Final He concentration due to ^{10}B transmut. after irradiation in	EUROFER97 (appm)	ADS2 (appm)	ADS3 (appm)	ADS4 (appm)
SPICE (13.6 dpa)	10.2	83.6	432.0	5580
ARBOR1 (22.4 dpa)	<1 [a,b]	6.2	43.3	
ARBOR2 (69.8 dpa)	<5 [a]	23.5	121.0	

[a] - estimation, [b] - EUROFER97 samples irradiated up to 32 dpa

Table 3.2.: Different heat treatments of EUROFER97 and boron doped alloys ADS2 to ADS4. Further information presents corresponding grain size (GS) and results from impact tests [23, 24].

	Heat treatment	GS [μm]	USE [J]	DBTT [°C]	σ_{ys} [MPa]
EUROFER97 ANL	980 °C/0.5 h + 760 °C/1.5 h	16	9.84	-81	543
EUROFER97 WB	1040 °C/0.5 h + 760 °C/1.5 h	21.4	9.84	-91	486
ADS2	1040 °C/0.5 h + 760 °C/1.5 h		8.81	-74	440
ADS3	1040 °C/0.5 h + 760 °C/1.5 h		8.92	-100	441
ADS4	1040 °C/0.5 h + 760 °C/1.5 h		5.5	-12	460

* dynamical yield stress at 100 °C

low helium concentration, but the transmutation rate is nearly constant even for the high ARBOR2 irradiation time. For simulation purposes, conditions in AR-BOR, therefore, are more appropriate for model development and validation since irradiation parameters do not significantly change over the whole experiment.

Two different heat treatments had been performed on the above mentioned steels: while EUROFER97 ANL was left in the *as delivered* state, EUROFER97 WB and all ADS model alloys were austenitized at higher temperatures, yielding slightly changed microstructural and mechanical properties as shown in Tab. 3.2. For further discussion of mechanical properties of boron doped steels, EUROFER97 WB will be used as reference to provide the same microstructural configuration.

3.4.2. Mechanical properties

Results obtained by impact tests [23, 79] on SPICE specimens are presented in Fig. 3.7. In Fig. 3.7a, unirradiated EUROFER97 WB and boron doped model alloys are compared. As one can observe, EUROFER97 WB, ADS2 and ADS3 show similar evolution of impact energies with test temperature, which serves as proof that the addition of these amounts of boron does not change steel's microstructure substantially. ADS4, on the contrary, shows severe loss of ductility (Upper Shelf Energy (USE)) and shift of Ductile-to-Brittle-Transition Temperature (DBTT) in comparison with base EUROFER97 WB steel even in the unirradiated state. That is why ADS4 was not considered in further investigations to access helium effects. Fig. 3.7b, c and d present results of impact tests for EUROFER97 WB, ADS2 and ADS3, respectively, for the unirradiated material (open symbols) and irradiated to 16.3 dpa in SPICE at 250 °C (circles) and 450 °C (diamonds). For all three materials, the most deteriorating effect of neutron irradiation occurs at low irradiation temperatures between 250 and 300 °C. The impact curves determined for 450 °C irradiation show less embrittlement, i.e. higher USE and lower DBTT. While impact properties for EUROFER97 and ADS2 show only minor differences between unirradiated state and irradiation at 450 °C, embrittlement of ADS3 remains clearly visible even at higher irradiation temperatures.

Hardening behavior studied by tensile tests was analyzed in [22]. Figs. 2.7 and 3.8a and b show stress-strain diagrams derived from SPICE for EUROFER97 WB, ADS2 and ADS3, respectively. Unirradiated samples were tested at different temperatures, while results from irradiated specimens were obtained by testing at the same temperature at which the irradiation took place. In the unirradiated state, all three materials show similar tensile behavior with comparable Ultimate Tensile Strength (UTS), but elongation is reduced for ADS2 and ADS3. After irradiation, the highest increase in the yield strength occurs after irradiation at 300 °C for all alloys. Again in this state, elongation for boron doped steels is reduced compared to EUROFER97. Differences between ADS2 and ADS3 were ascribed to helium effects, since ADS3 with a 5 times higher helium concentration showed a higher increase in yield strength, which was also confirmed by determination of a 10% higher hardness [22].

In Fig. 3.9a hardening and embrittlement results from SPICE, ARBOR1 and ARBOR2 are shown. Obviously, hardening evolution derived from the tensile and impact test seems to behave differently for SPICE samples. It has to be mentioned that the available specimen amount was very small due to the limited space in the experimental irradiation facilities. Therefore, statistics are quite poor. For tensile tests in SPICE, two specimens were tested under the same conditions, with error bars depicted in the diagram. A comparison to the hardening and embrittlement results from ARBOR1 [80, 81] and ARBOR2 [82] is shown. Influences of irradiation lead to a similar curve evolution for ARBOR1 and ARBOR2, while

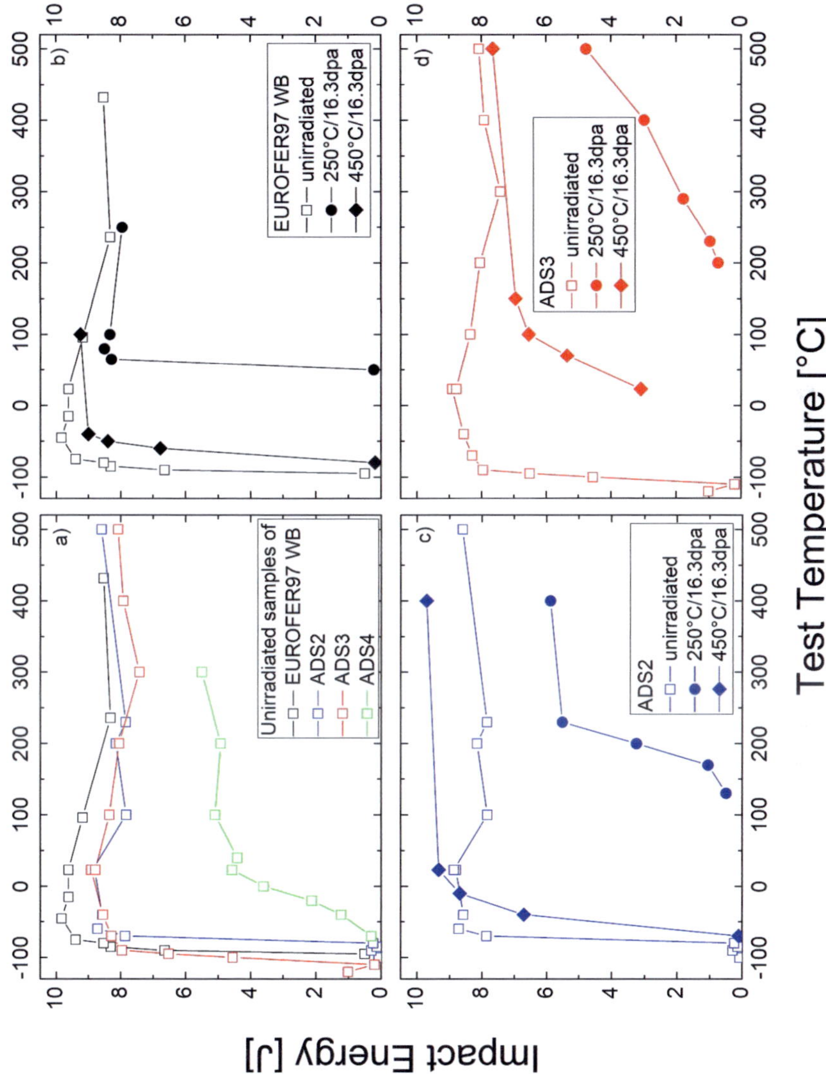

Figure 3.7.: Results of impact test measurements [23, 79] on a) EUROFER97 WB, ADS2 and ADS3 in the unirradiated state. Parts b) to d) show comparisons for each material after SPICE irradiation to 16.3 dpa at 250 and 450 °C. Lines are only to guide the eye.

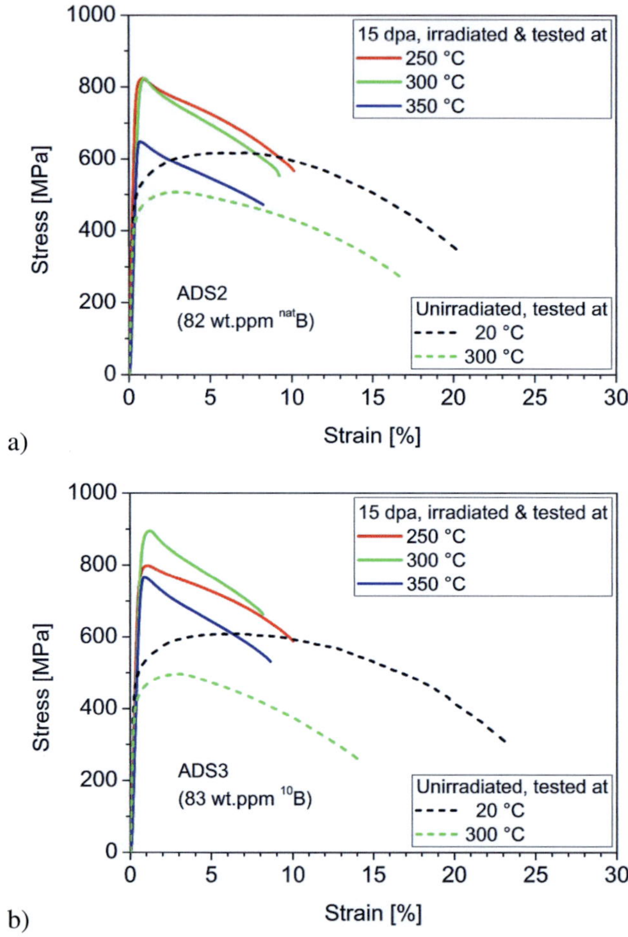

Figure 3.8.: Tensile tests of a) ADS2 and b) ADS3 specimens before and after irradiation to a dose of 15 dpa at different temperatures [22]. Behavior of both materials is qualitatively similar to EUROFER97 WB in Fig. 2.7, showing an increase in yield strength and loss of ductility under irradiation, with highest effects at 300 °C.

a higher dose and therefore higher helium amount and dislocation loop density for ARBOR2 specimens cause a curve shift to higher values of hardening and embrittlement. Since compositions and heat treatments of EUROFER97 (WB), ADS2 and ADS3 are widely similar, changes in mechanical properties have to stem from different helium concentrations within one irradiation experiment. Extra hardening and embrittlement (i.e. $\Delta DBTT_{ADS2/ADS3} - \Delta DBTT_{EUROFER97}$, analog for hardening) normalized to the maximum value is shown in Fig. 3.9b depending on the helium amount generated by boron transmutation. Negative values for extra hardening from SPICE tensile tests are omitted. A least square fit is applied on the basis of $y = Ax^b$ with exponent b represented by 0.33 and 0.5 for extra embrittlement and hardening, respectively. While extra $\Delta DBTT$ is well described by the fitting curve, values for extra $\Delta\sigma$ significantly scatter especially when the helium concentration is low.

3.4.3. Microstructural investigations

Microstructural investigations provide the basis for model development and simulations, which in our case rely on experiments using the boron doping technique (as described in Section 2.5). Before studying irradiated specimens, knowledge of boron distribution in the doped steel matrix is of great importance. Particularly, boron segregation to grain boundaries is a concern because boron accumulation would lead to high helium generation in these regions. Apparently, helium would deteriorate mechanical properties to a greater extent, as if it was generated homogeneously in the steel matrix, and firstly needed to diffuse to build clusters or bubbles. In a second step of course, the determination of experimental helium bubble size distributions is necessary, giving the possibility to adapt and verify the model. Microstructural investigations were performed by Auger Electron Spectrometry (AES) and Transmission Electron Microscopy (TEM).

AES investigations [84] were carried out with a PHI 680 Field Emission Scanning Auger Nanoprobe from Physical Electronics, providing a depth resolution of 2-3 nm and a lateral resolution of 20 nm. Samples were polished to 3 μm. After locating an area of interest by using the nanoprobe's Scanning Electron Microscope (SEM), surface cleaning using an argon ion beam (2 kV, 500 nA) was done by ablating 100 nm of the surface. AES analysis was performed using an accelerating voltage of 10 kV and a target current of 20 nA.

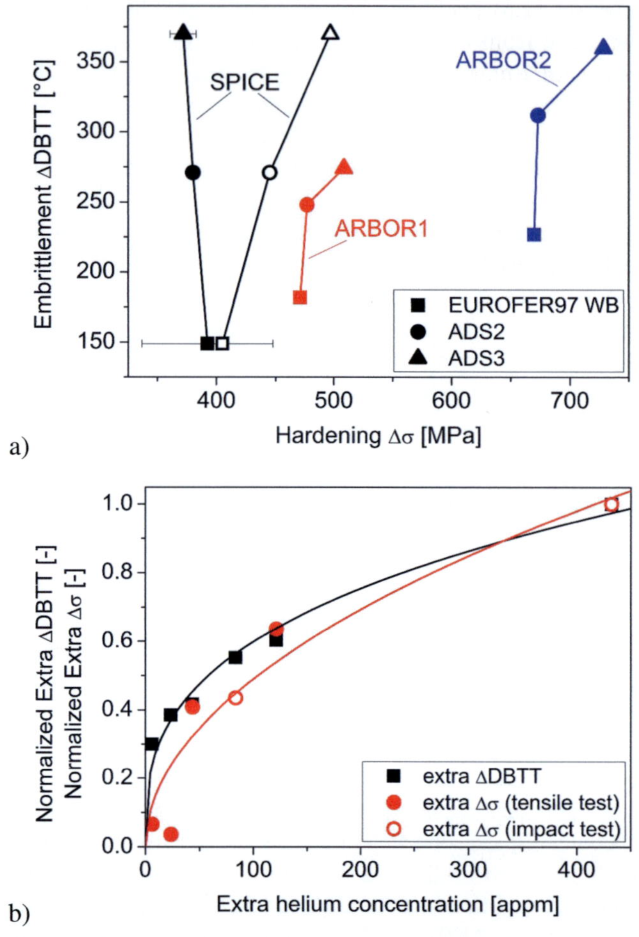

Figure 3.9.: a) Hardening and embrittlement after SPICE [22, 23, 79], ARBOR1 [80, 81] and ARBOR2 [82] experiments (results summarized in Tabs. A.1 and A.2). Hardening was determined by the difference in yield strength from tensile tests (full symbols), and by calculation of dynamic yield stress from impact tests [83] (open symbols). Error bars are given for tensile results after SPICE irradiation. b) Normalized extra hardening and embrittlement as a function of extra helium concentration due to boron transmutation. Negative data points from SPICE tensile tests were omitted, extra hardening and embrittlement were fitted by a power law.

Investigations by TEM [14, 31] were performed on non-irradiated as well as irradiated material from impact tested specimens. Samples were prepared by cutting slices of 150 μm, and subsequently thinning by electrolytic polishing until electron beam transparency was achieved. To reduce radioactivity and magnetism, whereof the latter influences the electron beam during investigations, small discs of 1 mm were punched out and put into a foldable copper net, which itself was placed into a double tilt specimen holder for the TEM investigations. Two microscopes of the same type were available, one of them operating in the Hot Cells of Fusion Materials Laboratory (FML) of KIT for investigating activated samples. In both cases, a FEI Tecnai G^2 F20 microscope was used, accelerating electrons by a voltage of 200 kV, and equipped with a post-column Gatan Image Filter (GIF) for Electron Filtered Transmission Electron Microscopy (EFTEM) measurements. An energy slit of 20 eV was chosen, which allowed improvement of contrast by removing the effects of inelastic scattered electrons due to zero-loss filtering. For detecting helium, Scanning Transmission Electron Microscopy (STEM) mode was used to perform Electron Energy Loss Spectroscopy (EELS) measurements on chosen bubbles.

Finally, comparisons with further quantitative microstructural investigations on helium bubbles published in literature are presented.

3.4.3.1. Boron distribution in alloyed steels

Boron distribution in unirradiated ADS2 and ADS3 was investigated by AES and TEM. The advantage of AES is a low detection limit of 1 atomic percent together with a lateral resolution of 20 nm for light elements like boron, while Energy Dispersive X-ray Spectroscopy (EDX) is typically more suitable for heavier elements. Nevertheless, given that boron concentration in ADS2 and ADS3 is about 0.04 atomic percent, homogeneously distributed boron is not detectable in the steel matrix. Since only boron accumulation would influence the helium generation and distribution, not measuring boron would imply that local boron concentration does not exceed 1 at.% throughout the whole specimen.

Results of AES investigations on an unirradiated ADS2 specimen are shown in Fig. 3.10. A typical Scanning Electron Microscopy (SEM) image is given in Fig. 3.10a. The red cross indicates the spot where the AES element analysis in Fig. 3.10b was performed. AES element maps were imaged for boron and nitrogen in Fig. 3.10c and d, respectively. Besides spherical precipitates, which were

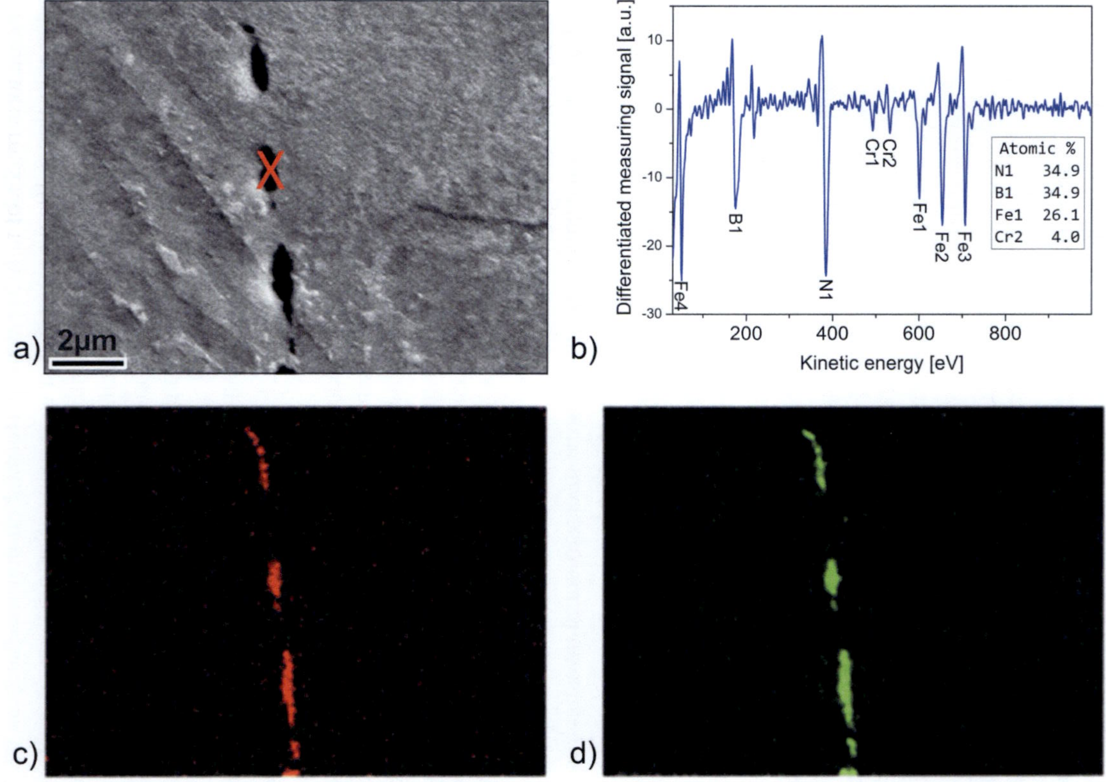

Figure 3.10.: AES analysis on unirradiated ADS2. In a) a SEM micrograph of the studied area is shown, the red cross indicates the spot where the AES elemental analysis (b) was performed. Element maps for c) boron and d) nitrogen are given.

identified as carbides of tantalum, tungsten or vanadium, elongated precipitates as shown in Fig. 3.10a were observed, and identified as boron nitride by AES element analysis. BN precipitates are stretched all in one direction, most likely the rolling direction, but otherwise are statistically distributed in the matrix with no preferential location at grain boundaries or other defects. For comparison, an ADS3 specimen was also investigated. It was shown that both times boron behaved the same way and was found as BN in the matrix. Measurements of BN precipitate densities were not performed, since precipitates were found infrequently, and the SEM magnification did not allow for good statistics. It has to be emphasized that due to the limited resolution monolayers of boron may exist at grain boundaries or other defects, and are not detectable by this technique. Nevertheless, even if boron precipitates during the casting process, BN seems to be distributed throughout the matrix, which will lead to nearly homogeneous helium generation.

TEM analysis [14] on ADS3 revealed results comparable to AES. Only a few precipitates containing boron were found by using EFTEM, and were identified as BN as shown in Fig. 3.11. The rotated central image demonstrates that BN is hardly visible in usual TEM bright field mode. Within an investigated volume of $1.748 \times 1.748 \times 0.105\,\mu m^3$, the shown BN particle was the only one found. By using EELS the number of boron atoms contained in this precipitate was estimated to be 4.7×10^6, while the whole volume close to the precipitate should cover 12.7×10^6 boron atoms based on the steel composition. As mentioned for the AES measurements, statistics are quite poor, since there exist also high uncertainties in precipitate shape, sample thickness and EELS energy window. Nevertheless, investigations confirmed that only a small amount of alloyed boron forms BN precipitates when considering the whole sample area, which are distributed randomly throughout the matrix.

3.4.3.2. TEM investigation on helium bubbles

Providing an experimental basis for the developed model, TEM investigations on ^{10}B doped ADS3 after ARBOR1 irradiation (22.4 dpa, 338 °C) were performed [14, 31]. The objective of that work was to determine helium bubble sizes and corresponding bubble densities. For this purpose, four different areas with a total size of $1.99\,\mu m^2$ on a ADS3 TEM sample were investigated, by counting and sizing all bubbles within this sample volume of $0.196\,\mu m^3$. Areas were chosen

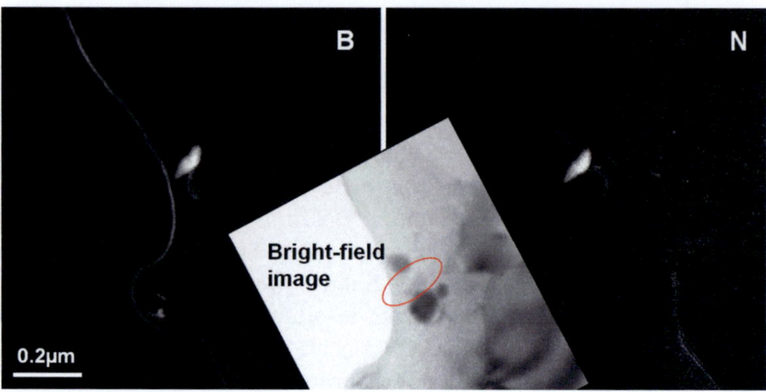

Figure 3.11.: EFTEM analysis on unirradiated ADS3 [14]. Comparable to AES results, precipitates of boron nitride could be identified, distributed randomly throughout the matrix. By EELS measurements it was proved that a low fraction of existing boron precipitates as BN.

because their images showed few interfering diffraction artifacts, and helium bubbles could be seen clearly. Exemplarily, the TEM micrograph in Fig. 3.12a shows a typical microstructure which was analyzed and measured. In this bright field image, helium bubbles appear as bright spots surrounded by a dark fresnel fringe in under-focus condition (-0.12 µm), while the contrast reverses for an over-focused electron beam. Since voids show the same diffraction contrast behavior, the existence of helium filled cavities was verified by EELS measurements [31]. The analyzed bubble, 4 nm of diameter, is shown in the TEM micrograph in Fig. 3.12b, together with the detected EELS spectrum. The energy loss Plasmon peak of the matrix material between 20 and 30 eV was fitted by a Gaussian curve (dashed line). Subtracting the fit from the measured curve, the inlay in Fig. 3.12b shows the resulting peak at 22.6 eV, which was assigned to the 1s-2p transition of helium [26]. Due to a shift of the peak when compared to the response of free helium (21.218 eV), the helium density inside the bubble could be estimated to 28 atoms per cubic nanometer [31]. Based on the lattice parameter of Fe-10% Cr, this density conforms to a helium-to-vacancy ratio of about 0.33.

Altogether, bubbles from sizes of around 0.5 nm to 13.6 nm were observed, besides a few exceptions all had a spherical shape. Bubbles are distributed homogeneously in the steel and showed no preferential decoration of grain

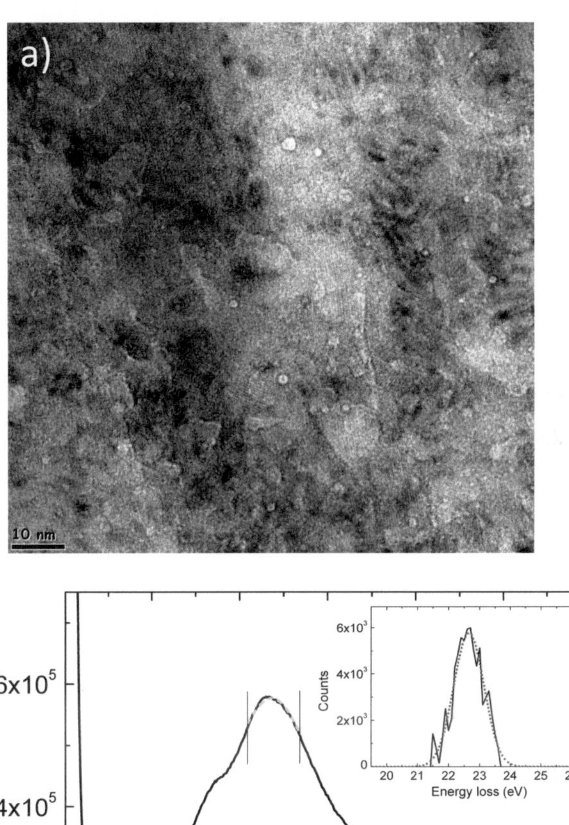

Figure 3.12.: a) TEM micrograph [14] showing typical distribution of helium bubbles in ADS3 after ARBOR1 irradiation. b) Helium identification by EELS [31] was performed on a 4 nm sized bubble (see inlay). The shift of the energy loss peak of helium (inserted diagram) can be attributed to a helium density inside the bubble of 28 atoms per cubic nanometer.

boundaries or line dislocations. It has to be emphasized that the smaller the bubbles the more difficult it is to detect them, because the diffraction contrast of the bubble is superposed by other effects, e.g. surface effects from sample preparation and TEM resolution. Assessing the existing TEM images, it seemed reasonable to identify and measure helium bubbles with sizes larger than 0.75 nm in this study, even though uncertainties are still high for such small bubbles.

Resulting histograms with bin size of 0.25 nm for each inspected area are shown in Fig. 3.13. Areas a) to d) were evaluated separately, inlays of Fig. 3.13a-d present the number of counted bubbles, the local total bubble density N and mean diameter $\langle d \rangle$. Size histograms show comparable results for all four areas. Size distributions have similar shapes, with peak sizes at small diameters between 1.1 and 1.6 nm. Larger bubbles above 4 nm are rarely observable, and statistics for these bubbles are bad, since sometimes only one bubble was found within the histograms' bin size of 0.25 nm. On the other hand, as mentioned earlier, small bubbles of sizes below 1 nm have a higher possibility of not being detected. Therefore one can argue whether a peak really exists in the diagrams, or whether a large amount of small bubbles exist with sizes below 1 nm, which are just not being visually resolved. Fig. 3.13b–d could be interpreted in the latter way, while the size distribution in Fig. 3.13a supports the existence of a real peak. In any case, most bubbles observed (97%) are smaller than 3 nm. Total bubble densities differ at most by a factor of 5, but other regions, which were not analyzed, may have a lower or higher density of helium bubbles.

3.5. Further microstructural investigations on helium bubbles

Few microstructural investigations on ferritic steels have been published so far providing quantitative analysis on helium bubbles.

In [85], ferritic Fe-14%Cr was investigated after dual-beam irradiation of 24.18 MeV Fe^{8+} and 1.7 MeV He^+ ions in the Joint Accelerators for Nanosciences and NUclear Simulation (JANNUS) facility [86]. Irradiation temperature was 425 °C, damage doses between 10 and 40 dpa and helium production rates up to 100 appm/dpa were achieved, depending on the implantation depth. Unfortunately, it is not described from which depth the TEM sample was prepared, and therefore

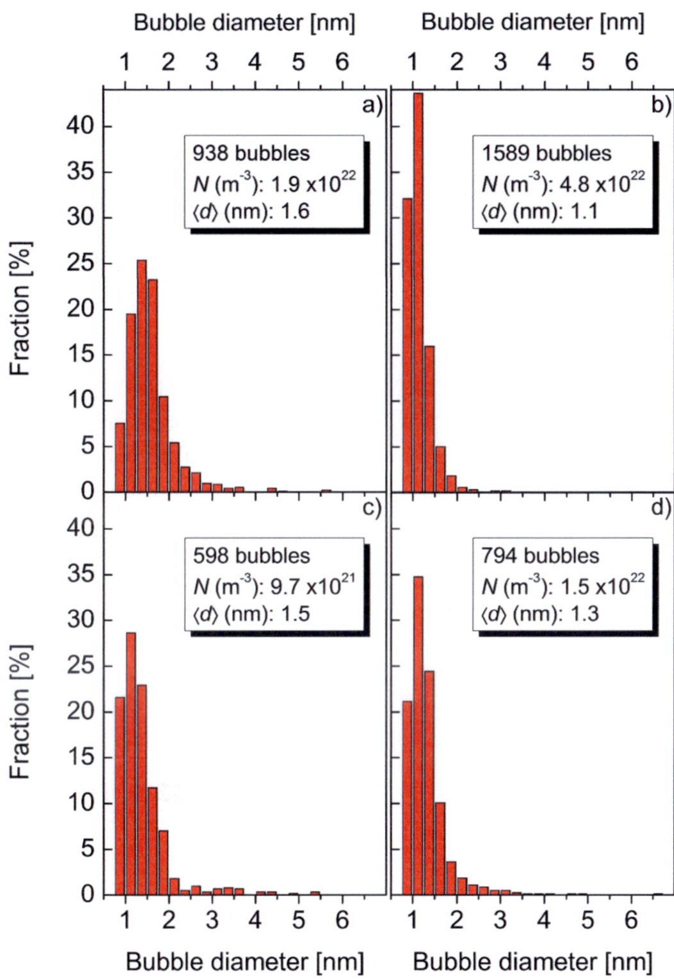

Figure 3.13.: Results of TEM investigation on irradiated ADS3 [14, 31] from AR-BOR1. Four different areas (a–d) were analyzed concerning helium bubble number and sizes. Histograms of size measurements are given, together with total bubble density N and mean diameter $\langle d \rangle$ calculated for each area.

the exact irradiation conditions are not known. Nevertheless, results show bubbles with diameters from 0.9 to 15 nm and a total bubble density of about 1.7×10^{22} m^{-3}. A high density of small bubbles is observed, and a few larger bubbles. Therefore, the bubble size distribution was divided into two distributions, one for bubbles up to 4.25 nm and the other for larger voids, presenting a bimodal size distribution. The cut-off diameter was chosen from the histogram and corresponds well with the critical cavity size derived from theoretical considerations for the onset of unstable bias driven void growth (see Section 3.1.2). Hence bubble size distribution shows a total density of 1.5×10^{22} m^{-3} and a mean size $\langle d \rangle$ of 2.1 nm, while for voids the total density is 1.9×10^{21} m^{-3} with $\langle d \rangle = 6.9$ nm. The amount of cavities larger than 4.25 nm is consequently 11% of the total quantity.

In [87], a high-Ta variant of the RAFM steel F82H (F82H-mod.3) was subjected to neutron irradiation in the High Flux Isotope Reactor (HFIR) at a temperature of 500 °C to a dose of 9 dpa. Helium was produced by a NiAl implantation layer of varying thicknesses deposited on the sample surface, achieving total helium concentrations of 190 and 380 appm. Measurements yield bubble densities of 3.9×10^{22} m^{-3} and 5.3×10^{22} m^{-3} and mean bubble diameters of 1.4 nm and 2 nm for the lower and higher helium amounts, respectively. In both cases, most of the bubbles observed were nanometer sized, but single larger cavities appeared, and a bimodal size distribution was assumed. Analyzing the histograms, one can say that in both size distributions 2% of the counted and measured cavities were larger than 5 nm.

3.6. Interaction of displacement cascades with helium bubbles

Neutron irradiation not only generates new helium by (n,α)-reactions, but moreover through collisions with iron lattice atoms their kinetic energy is in parts transferred to so-called primary knock-on atoms (PKAs). They are responsible for the creation of displacement cascades which interact within their zone of influence with existing helium bubbles in the matrix. For studying these effects, a ferritic Fe-12% Cr model alloy was exposed to irradiation with 100 keV He$^+$ at 973 K until 600 appm helium was implanted [42]. The microstructure was investigated by TEM and compared to the condition after a subsequent irradiation with 300 keV Fe^+ to 30 dpa at 573 K. As a result, helium resolution from bubbles was observed, in Fig. 3.14a halos of small helium bubbles surrounding larger ones can be observed. It was stated in [42], that helium resolution strongly depends on bubble size and its relation to the cascade volume:

- Small clusters (few atoms) completely dissolve.

- Bubbles of intermediate size of a few nanometers lose a substantial amount of their atoms in the cascade's hot spike phase, but resolved helium may be re-caught by the initial bubble, or re-nucleate into new bubbles.

- Large bubbles resolve a few atoms into the surrounding matrix, but will re-catch most of them after the cascade wears off.

Quantification of these theoretical considerations was done by means of MD simulations for specific PKA and bubble characteristics and configurations [88, 89, 90]. Calculation basis was the iron bcc lattice, PKAs were placed in distances to the bubble between 2 and 10 times the lattice constant a_{Fe}. In [88] simulations were performed for PKA energies, E_{PKA} of 5, 10 and 15 keV with helium-to-vacancy ratio x_{He} of unity and at a temperature of 500 K. PKA impulses were directed towards the bubble center along $\langle 135 \rangle$ crystallographic orientation. Simulation boxes of sizes $46 \times 45 \times 47\ a_{Fe}^3$, $54 \times 53 \times 52\ a_{Fe}^3$ and $59 \times 58 \times 60\ a_{Fe}^3$ were used for PKA energies of 5, 10 and 15 keV, respectively, providing periodic boundary conditions and using n-body Fe-Fe potential from [91], pairwise Fe-He potential from [92], and He-He pair potential from [93]. It has to be noted that statistics due to a limited number of simulation runs (2–6) are weak. Mean calculation results are shown in Fig. 3.14b and compared to simulations performed in [89, 90], where, amongst others, results were presented for $E_{PKA} = 5\,keV$, $T = 600\,K$ and varying x_{He}.

In Fig. 3.14b, complete dissolution of small clusters of two atoms can be observed, while the percentage of dissolved helium atoms from clusters decreases with increasing cluster size. For a nanometer sized bubble (40 atoms) both simulations show a resolution of about 2% of the contained helium atoms for $x_{He} = 1$, and 5% for $x_{He} = 0.25$ [89]. For a bubble with a diameter of 2 nm (339 atoms), 0.8% of the helium atoms dissolve when the bubble is hit by a 5 keV cascade, while effects of the helium-to-vacancy ratio can not be distinguished. The theoretical considerations in [42], that displacement cascades induce a bubble size dependent helium resolution most pronounced for small clusters, was thus verified by the presented MD simulations.

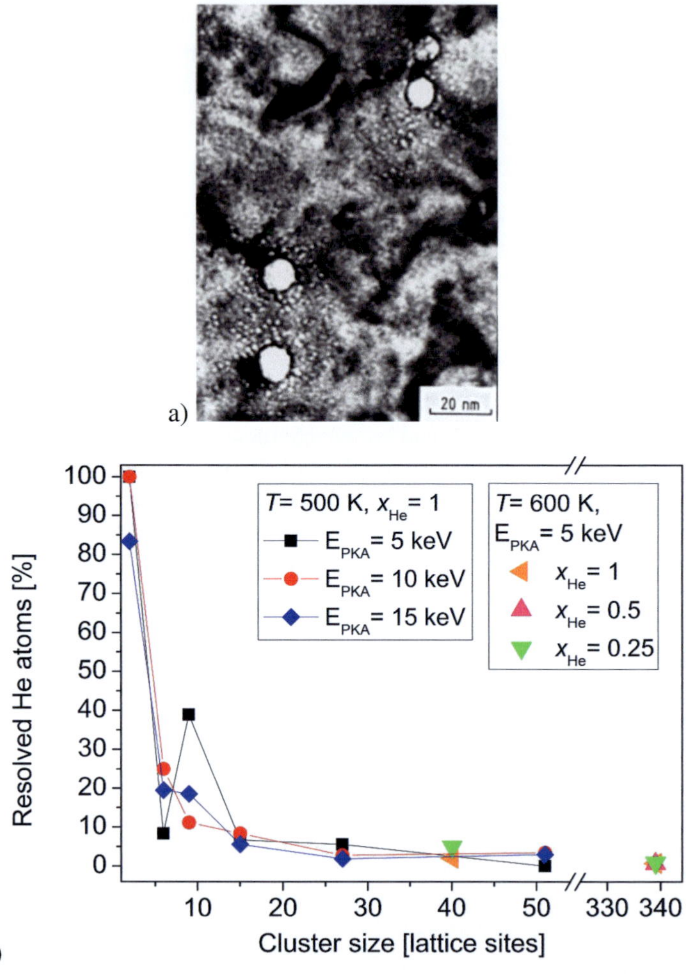

Figure 3.14.: a) TEM micrograph showing helium resolution from bubbles in ferritic Fe-12% Cr model alloy formed under $100\,keV$ He^+ implantation to $600\,appm$ at $973\,K$, and subsequent $300\,keV$ Fe^+ irradiation to $30\,dpa$ at $573\,K$ [42]. Halos of small bubbles generated under Fe^+ irradiation surrounding larger ones from prior He^+ implementation are visible. b) MD simulations performed at $500\,K$ [88] and $600\,K$ [89, 90] providing helium resolution from bubbles due to cascade-bubble interaction for different simulation conditions in α-iron. Helium resolution is more pronounced for small clusters, while for larger bubbles the resolved helium fraction decreases to about 0.8% at a bubble size of 339 lattice sites.

4. Model description

The developed model presented here uses Rate Theory to describe helium bubble nucleation and growth. The aim of this model is to allow the simulation of helium bubble growth up to times comparable to the duration of irradiation experiments and relevant for fusion. Under certain assumptions, Rate Theory provides the possibility of simulating time periods up to years keeping the calculation times within acceptable ranges. Input parameters are chosen with care, and derived from physically based literature results of *ab initio*, Molecular Dynamics (MD) or Monte Carlo (MC) methods.

4.1. Basic equations

Generally, a monomer M is captured or emitted by a cluster M_i, which contains i particles. This process takes place with a certain probability expressed by kinetic rate coefficients k_i for capturing and g_{i+1} for emitting of a monomer:

$$M + M_i \; \underset{g_{i+1}}{\overset{k_i}{\rightleftarrows}} \; M_{i+1}. \tag{4.1}$$

Fig. 4.1 presents the underlying scheme of the model, describing the evolution possibilities of a bubble with size i.

Clustering of helium atoms in solids cannot be easily described due to the strong binding of helium to several microstructural defects, e.g. vacancies or voids, grain boundaries, etc. From energetical considerations helium prefers to occupy a substitutional lattice site and thus forms helium-vacancy clusters. Therefore the role of vacancies and the diffusion of different helium-vacancy clusters also has to be considered when describing nucleation and growth of helium bubbles. In spite of extensive theoretical and experimental investigations (see Paragraph 3.2.1), there

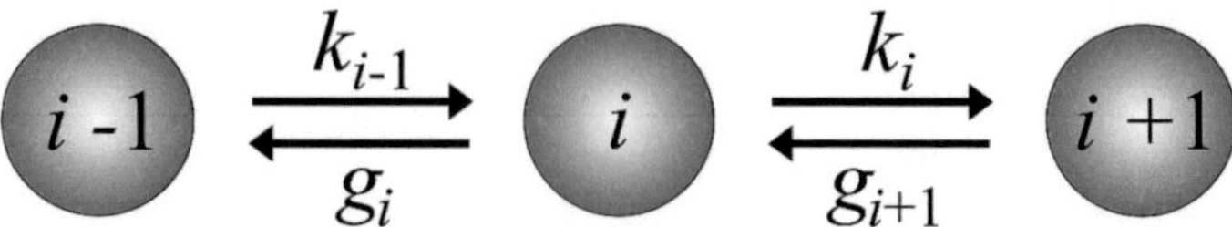

Figure 4.1.: Basic scheme of the model. A bubble with size i is allowed to grow or shrink by one helium atom (monomer). Capturing rates, k_i and emission rates, g_i describe the corresponding probabilities, respectively.

still exists no consensus on the dominant helium diffusion mechanism, which is expected to show a strong temperature dependence. Moreover, there exists only rare and partly inconsistent information about the helium density (helium-to-vacancy ratio) in a bubble [26, 61]. The current calculations are performed for a helium-to-vacancy ratio x_{He} of 1, 0.5 and 0.25, which is parametrically adjustable at the beginning but constant for all bubble sizes within one simulation. The values of x_{He} are justified by the following considerations:

1. MD simulations on a 2 nm sized helium bubble in iron [71] showed a steep helium pressure increase from 8 to 27 GPa at 300 °C, when x_{He} was altered from 1 to 2. At higher x_{He} the iron lattice even starts to deform plastically. Therefore, values for x_{He} larger than 1 were not considered.

2. MD and Molecular Static (MS) calculations in [70] yield the lowest energy configuration (formation energies) of helium bubbles for the helium-to-vacancy ratio of approximately unity.

3. Based on TEM investigations on irradiated ADS3 specimens, elemental analysis by Electron Energy Loss Spectroscopy (EELS) on helium bubbles was performed [31]. Results on one of the largest bubbles found (4 nm) showed that the helium density inside the bubble corresponds to a helium-to-vacancy ratio of ~ 0.33. These measurements also indicate that bubble evolution in the considered irradiation experiments did not reach an unstable void growth regime. Therefore, the range of x_{He} investigated here was fixed from 0.25 to 1.

Considering only single mobile defects in the matrix, the change in cluster concentrations $C_i(t)$ of different sizes i with time can be expressed by the following set of differential equations:

$$\frac{\partial C_1(t)}{\partial t} = G_{\text{He}}(t) - L_{\text{sinks}}(t) - \sum_i k_i C_1(t) C_i(t) + \sum_i g_i C_i(t) \qquad (4.2)$$

and (in accordance with Fig. 4.1)

$$\frac{\partial C_i(t)}{\partial t} = -\left[k_i C_1(t) + g_i\right] C_i(t) + g_{i+1} C_{i+1}(t) + k_{i-1} C_1(t) C_{i-1}(t), \qquad (4.3)$$

both valid for $i \geq 2$. Eq. (4.2) describes the change of the monomer concentration $C_1(t)$ depending on the helium generation rate $G_{\text{He}}(t)$, the loss rate to sinks $L_{\text{sinks}}(t)$ and the net helium monomer amount of catching and emitting of a monomer by all clusters. G_{He} will be defined later, when specific simulation conditions are discussed in Section 6.1, and L_{sinks} is described in Section 4.6. While Eq. (4.2) provides the left boundary condition for the numerical calculation carried out with a Fortran code, the first term of the Master Eq. (4.3) describes the decrease of the concentration of clusters with size i due to growing and shrinking, while the second and third term lead to an increase in cluster concentration by shrinking of larger clusters and growth of smaller clusters, respectively.

4.2. Cluster radius

The iron lattice features a body-centered cubic (bcc) structure, therefore an elementary cell with lattice constant a_{Fe} owns two atoms on a regular lattice site. The volume occupied by an atom, Ω, can be expressed by

$$\Omega = \frac{a_{\text{Fe}}^3}{2}. \qquad (4.4)$$

Monomers of substitutional helium own a volume V_1, which depends on the helium-to-vacancy ratio x_{He}, given by

$$V_1 = \frac{\Omega}{x_{\text{He}}}. \qquad (4.5)$$

Helium clusters are assumed to be spherical. Their volume, V_i containing i monomers is defined by

$$V_i = \frac{4}{3}\pi R_i^3 = iV_1. \tag{4.6}$$

By this means, and using Eq. 4.5, the volume of a bubble of size i is expressed by summing up the volumes of each monomer. The resulting cluster radius R_i is given by

$$R_i = \sqrt[3]{\frac{3\Omega i}{4\pi x_{\text{He}}}} = b\sqrt[3]{i}, \tag{4.7}$$

with parameter b defined as an equivalent monomer radius. In [68] temperature dependent lattice constants are calculated for 10 percent chromium alloyed iron:

$$\frac{a_{\text{Fe10Cr}}}{\text{Å}} = 2.856 + 3.685 \times 10^{-5}\frac{T}{\text{K}}, \tag{4.8}$$

and hence will be used for simulations in this work.

4.3. Effective helium diffusivity

Taking the general thoughts from Section 3.2.1 into account, the effective helium diffusivity $D_{\text{He}}^{\text{eff}}$ used in this model is taken as a fitting parameter close to values of the helium-divacancy diffusion coefficient [67] shown in Fig. 3.5 .

4.4. Kinetic rate coefficients

4.4.1. Capture rate coefficient k_i

To solve the Master Eq. (4.3) the kinetic rate coefficients have to be defined. Basically, monomer capturing by a cluster involves diffusion of monomers to clusters and an attaching reaction. Therefore, capture rates can either be distinguished by diffusion ($k_{i,\text{diff}}$) or reaction limited processes ($k_{i,\text{react}}$). Considering the diffusion controlled process first, the capture rate is described by the effective diffusion coefficient and by geometrical parameters A_i for active surface and by effective radius R_{eff}. In the region within this radius the cluster is capable of capturing a monomer.

$$k_{i,\text{diff}} = \frac{A_i}{R_{\text{eff}}} D_{\text{He}}^{\text{eff}} \tag{4.9}$$

In this model helium clusters are assumed to be spherical. Therefore the geometrical parameters are adapted:

$$A_i = 4\pi R_i^2, \tag{4.10}$$

$$R_{\text{eff}} = R_i. \tag{4.11}$$

Inserting Eq. (4.10) and (4.11) in (4.9) yields the diffusion limited capture rate:

$$k_{i,\text{diff}} = 4\pi R_i D_{\text{He}}^{\text{eff}}. \tag{4.12}$$

The reaction limited process is characterized by an activation energy E_i for capturing a monomer which leads to a Boltzmann expression with Boltzmann's constant k_{B}. The kinetic rate further depends on the surface area of the cluster [94].

$$k_{i,\text{react}} = \frac{4\pi R_i^2}{a_{\text{Fe}}} D_{\text{He}}^{\text{eff}} \exp\left[-\frac{E_i}{k_{\text{B}} T}\right] \tag{4.13}$$

The diffusion and reaction limited processes can be described by one equation. Hence, it depends on the underlying cluster growth mechanism which process dominates the capture rate. Because the diffusion and reaction processes proceed sequentially, the corresponding time periods have to be added linearly. Conversely, by combining the reciprocals of the diffusion and reaction limited rates one obtains the following equation for the total capture rate k_i:

$$k_i = \frac{4\pi R_i D_{\text{He}}^{\text{eff}}}{1 + \frac{a_{\text{Fe}}}{R_i} \exp\left[\frac{E_i}{k_{\text{B}} T}\right]}. \tag{4.14}$$

The attachment barrier E_i is not known exactly, but it is supposed to be small with a value in the range of the interstitial migration energy of helium (0.06 eV). Generally, for simplification and reduction of unknown parameters, the capture rate is assumed to be diffusion limited in all performed simulations. Additionally, effects of an attachment barrier on simulation results are assessed within a parameter study taking E_i as a fitting value.

4.4.2. Emission rate coefficient g_{i+1}

The emission rate g_{i+1} of monomers from clusters is described under the assumption of steady state condition, which implies equality of monomer capturing and emitting probabilities:

$$g_{i+1} C_{i+1}^{\mathrm{eq}} = k_i C_1^{\mathrm{eq}} C_i^{\mathrm{eq}}. \tag{4.15}$$

The concentrations in Eq. (4.15) can be replaced by the atomic density of the iron lattice (expressed by the reciprocal of the atomic volume Ω) and atomic fractions x_i of the different defects i, i.e.

$$g_{i+1} = k_i \left[C_1 \frac{C_i}{C_{i+1}} \right]_{\mathrm{eq}} = k_i \Omega^{-1} \left[x_1 \frac{x_i}{x_{i+1}} \right]_{\mathrm{eq}}. \tag{4.16}$$

The relation of the fractions x_i can be expressed by the following equation analog to [95]:

$$\left[\frac{x_1 x_i}{x_{i+1}} \right]_{\mathrm{eq}} = \left[\frac{\theta_1 \theta_i}{\theta_{i+1}} \right]_{\mathrm{eq}} \cdot \exp \left[-\frac{G_1^{\mathrm{f}} + G_i^{\mathrm{f}} - G_{i+1}^{\mathrm{f}}}{k_{\mathrm{B}} T} \right]. \tag{4.17}$$

Therein, θ_i are the internal degrees of freedom for defects i, G_i^{f} the Gibbs free energies of formation of a defect cluster containing i monomers, herein after referred to as formation energies. The relation of the internal degrees of freedom are assumed to be 1 [96]. The equilibrium concentration of monomers is obtained by statistical thermodynamics [62]:

$$C_1^{\mathrm{eq}} = \Omega^{-1} \exp \left[-\frac{G_1^{\mathrm{f}}}{k_{\mathrm{B}} T} \right]. \tag{4.18}$$

Combination of Eqs. (4.15-4.18) yields the equation for the emission rate g_{i+1} as a function of the formation energy difference for size adjacent clusters:

$$g_{i+1} = k_i C_1^{\mathrm{eq}} \exp \left[\frac{G_{i+1}^{\mathrm{f}} - G_i^{\mathrm{f}}}{k_{\mathrm{B}} T} \right]. \tag{4.19}$$

As mentioned previously, formation energies can be obtained for different cluster sizes e.g. by MD calculations [70]. Nevertheless, the effort of calculation is quite high, that is why only formation energies for small clusters (less than 100 helium atoms) are available in the literature. In the following paragraph a macroscopic material constant, the surface energy γ_{SF} (of iron) [97, 98, 99, 100], is introduced, which is assumed to be independent of the cluster size i and which replaces the formation energies in the emission rate coefficient.

Cluster formation energy G_i^{f} and nucleation energy G_i^{n} are connected to each other as shown next:

$$G_i^{\mathrm{n}} = G_i^{\mathrm{f}} - iG_1^{\mathrm{f}}. \tag{4.20}$$

Following classical nucleation theory [60], the nucleation of a new phase affects the Gibbs free energy of the system in two opposing ways: the energy is lowered because of the nucleation of a new homogeneous volume while the creation of an interface between the two phases increases the Gibbs free energy.

$$G_i^{\mathrm{n}} = -G_1^{\mathrm{f}} i + \gamma_i i^{\frac{2}{3}} \tag{4.21}$$

The first term on the right-hand side of Eq. (4.21) shows the contribution of the volume, the second one gives the surface term depending on the surface energy per cluster atom γ_i. The differences of the nucleation energies of clusters with size i and $i+1$ ($\Delta i = 1$) can be expressed by the derivative of Eq. (4.21) with respect to i:

$$G_{i+1}^{\mathrm{n}} - G_i^{\mathrm{n}} \approx \frac{\mathrm{d}G_i^{\mathrm{n}}}{\mathrm{d}i}\Delta i = -G_1^{\mathrm{f}} + \frac{2}{3}\gamma_i i^{-\frac{1}{3}}. \tag{4.22}$$

Including Eqs. (4.18), (4.20), (4.22) into (4.19) yields the emission rate coefficient depending on the surface energy per cluster atom γ_i:

$$g_{i+1} = k_i C_1^{\mathrm{eq}} \exp\left[\frac{\frac{2}{3}\gamma_i i^{-\frac{1}{3}}}{k_{\mathrm{B}}T}\right]. \tag{4.23}$$

The surface term of Eq. (4.21) can be related to the surface energy γ_{SF}:

$$\gamma_i i^{\frac{2}{3}} = \gamma_{\mathrm{SF}} \cdot 4\pi R_i^2$$

$$\gamma_i = 4\pi R_i^2 i^{-\frac{2}{3}} \gamma_{\mathrm{SF}}. \tag{4.24}$$

Eqs. (4.4) and (4.24) are inserted into (4.23). Therefore, the emission rate coefficient can be expressed with the macroscopic material parameter γ_{SF} by the following equation:

$$g_{i+1} = k_i C_1^{\mathrm{eq}} \exp\left[\frac{2\Omega\gamma_{\mathrm{SF}}}{b\sqrt[3]{i} \cdot k_{\mathrm{B}}T}\right]. \tag{4.25}$$

It has to be emphasized that Eq. (4.25) yields a size dependent emission rate coefficient where smaller clusters have a higher tendency to emit a monomer and therefore are more likely to shrink than larger clusters leading to a coarsening mechanism similarly described by [61].

4.5. Helium solubility

For the description of the kinetic coefficients the solubility (or equilibrium concentration) C_1^{eq} of helium in the iron matrix must be obtained. In principle, helium atoms in solution and in the gas phase have to be in equilibrium. By equalizing their chemical potentials as shown in [101], helium solubility can be calculated. Although the use of the chemical potential in the real gas phase would be a standard approach for solubility calculation, we have used an ideal gas approximation for estimation of solubility which was a starting point for a parametric study of its influence on the bubble evolution. An approximation for an ideal gas is described in [102] and can be expressed by

$$C_1^{\mathrm{eq}} = X \frac{p_{\mathrm{He}}}{p_0} \exp\left[\frac{S_F}{k_B}\right] \exp\left[-\frac{G_1^{\mathrm{f}}}{k_B T}\right], \tag{4.26}$$

where X is the number of possible positions of helium atoms per number of unit cell atoms ($= 2$) of the iron lattice, p_{He} the helium bubble pressure and p_0 the standard pressure (in $\mathrm{N/m^2}$) as given by Eqs. (4.27) and (4.28), respectively, with helium mass m and Planck constant h [102]:

$$p_{\mathrm{He}} = \frac{2\,\gamma_{\mathrm{SF}}}{R_i}, \tag{4.27}$$

$$p_0 = \frac{(k_B T)^{\frac{5}{2}} (2\pi m)^{\frac{3}{2}}}{h^3}. \tag{4.28}$$

Eq. (4.26) is applicable to describe the equilibrium of free helium gas from bubbles with solvent helium in the solid at substitutional ($X = 1$) as well as interstitial position ($X = 6$), using the appropriate defect formation energies of $G_{1,\mathrm{sub}}^{\mathrm{f}} = 3.25\,\mathrm{eV}$ and $G_{1,\mathrm{int}}^{\mathrm{f}} = 5.25\,\mathrm{eV}$ [103], respectively. For the latter case, recent *ab initio* calculations of helium in bcc iron showed that the tetrahedral interstitial configuration is most stable [66]. The formation entropy S_F is calculated by the Sackur–Tetrode equation for ideal gases [104]:

$$S_F = k_B \, i \left[\frac{5}{2} + \ln \frac{(k_B T)^{\frac{5}{2}} (2\pi m)^{\frac{3}{2}}}{p_{He} h^3} \right].$$

(4.29)

Since solubility to interstitial position is much lower, substitutional solubility C_1^{eq} was calculated for the irradiation experiments' relevant temperatures, as shown in Tab. 6.2.

4.6. Loss rate at sinks

Single helium atoms may be caught by sinks and will not take part in further clustering in the matrix. Lattice defects like dislocations (DLs), interstitial loops and vacancy clusters, but also precipitates and grain boundaries (GBs), can act as sinks for helium atoms. In this paragraph a general expression is presented describing the loss rate L_{sinks} of monomers at sinks [105]:

$$L_{sinks} = D_{He}^{eff} C_1(t) \, k_{sinks}^2.$$

(4.30)

The loss rate is proportional to the diffusivity of the monomer and its concentration. The sink strength k_{sinks}^2 sums up all sink strengths of existing defects j in the matrix and can be expressed by

$$k_{sinks}^2 = \sum_j k_j^2.$$

(4.31)

Sink strengths for various defects can be derived from [69, 106, 107] and are shown in Tab. 4.1. For simplicity interstitial loops are considered spherical.

It is assumed that once helium atoms are trapped at sink sites they will not be emitted to the matrix any more. Nevertheless, the helium is not lost, but stored at the different sink sites. Therefore, depending on the sink strengths, the sinks themselves may act as nucleation sites for further helium clustering.

Table 4.1.: Sink strengths of different defects. Z_{He} is the bias factor, ρ_{disl} the dislocation density, R_{loop} the interstitial loop radius, C_{loop} the interstitial loop concentration, and R_{grain} is the grain radius.

Sink j	Sink strength k_j^2	
Line dislocation	$Z_{He}\,\rho_{disl}$	[69]
Interstitial loop	$4\pi R_{loop}C_{loop}$	[106]
Spherical grain boundary	$15/R_{grain}^2$	[107]

4.7. Cascade induced helium resolution from bubbles

The interaction of displacement cascades with existing bubbles was already described in Section 3.6, showing resolution of helium atoms from bubbles into the matrix. The effect is implemented into the basic model because the modification is assumed to show a noticeable influence on resulting bubble size distributions. Fig. 4.2 demonstrates how cascade induced helium resolution is implemented in the model. For a given simulation time t, the simulated bubble size distribution (grey) is recalculated before initiating the next time step Δt. Taking the red cross as an example, a size reduction of 50% is assumed (blue horizontal arrow). This is of course too large when comparing to results of MD simulations in Section 3.6, but is used for demonstration purposes here. Not all bubbles of that size are affected by cascades, just a certain fraction which is calculated by taking into account simulating time step Δt, damage dose rate and a correlation factor. The calculated concentration fraction is substracted from that bubble size and added to the concentration of bubbles with 50% of the size (see blue vertical arrows). The black shaded area symbolizes the released concentration of helium atoms due to size reduction, which is thus added to the monomer concentration solved in the matrix (black arrow, not to scale). In that way the whole size distribution is recalculated for each cluster size after each time step Δt.

The results from MD simulations shown in Fig. 3.14b are used to estimate the expected size reduction due to neutron induced primary knock-on atoms (PKAs). While at first the model is verified to work for a fixed fraction of resolved helium atoms of 5%, a bubble size dependent helium resolution is also implemented using

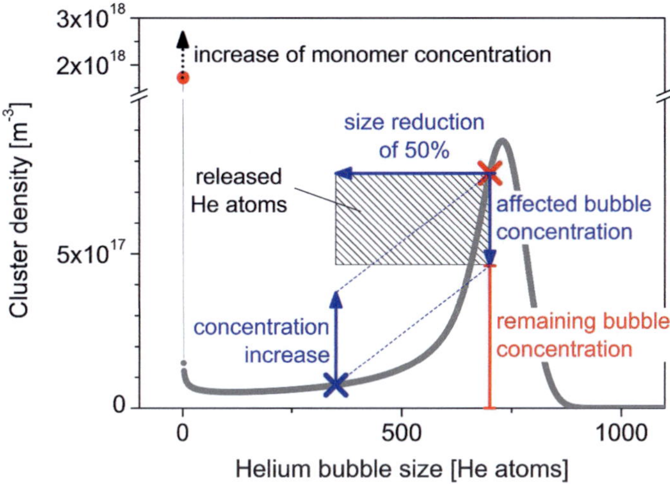

Figure 4.2.: Demonstration how cascade induced helium resolution from bubbles is implemented into the model. For a fixed size reduction the affected helium concentration is calculated, substracted and added to the target bubble size. Released helium increases monomer concentration.

a fitting curve of the MD data. Calculation of the affected cluster concentration, C_i^{aff} of cluster size i is done by using the relation

$$C_i^{\mathrm{aff}}(t) = X_{\mathrm{corr}}\, G_{\mathrm{dpa}}\, \Delta t\, C_i(t), \qquad (4.32)$$

where X_{corr} is the correlation factor. For the time being, X_{corr} is taken as a fitting parameter to visualize the effect of the model modification. Later on, a discrete description has to be found using

- PKA generation rates causing displacement cascades by impacting neutrons with characteristic energy distribution,

- characteristic cascade volumes affecting bubbles within their range.

For $X_{\mathrm{corr}} = 50$ and a constant size reduction of 5%, in Appendix B.4 the numerical code is presented which was implemented into the main program.

4.8. Clustering at GB sinks

The basic three-dimensional (3-D) model is modified to additionally describe helium bubble nucleation and growth at sink sites, in this case GBs. The motivation for describing this is partly based on Transmission Electron Microscopy (TEM) measurements performed in [29], where helium bubbles were observed under high temperature irradiation decorating sink sites (in this case mainly DLs). Since sinks may catch a large amount of helium during irradiation, this concentration does not take part in clustering in the 3-D model and is omitted from the calculations. Obviously, sink capacity for helium in solution is limited, and large helium amounts will lead to bubble nucleation at these sites. Although bubble nucleation at sinks would *per se* be attributed to a heterogeneous type, the two-dimensional (2-D) modification of the model describes homogeneous nucleation considering the whole area of GBs.

Fig. 4.3 shows the main idea of the 2-D model. Grains are assumed to have a spherical shape. The corresponding grain surfaces are then taken as a uniform plane where homogeneous nucleation takes place. Although GB triple points are assumed to be a highly disordered region providing extended space for solving helium, their effects are ignored because of their low frequency in comparison to the whole GB area. The area of GBs is defined as a plane of a thickness of one atomic layer. Evolving bubbles have their middle point within the plane, intersecting the GB area as a circle, and interacting with the plane through their circumference.

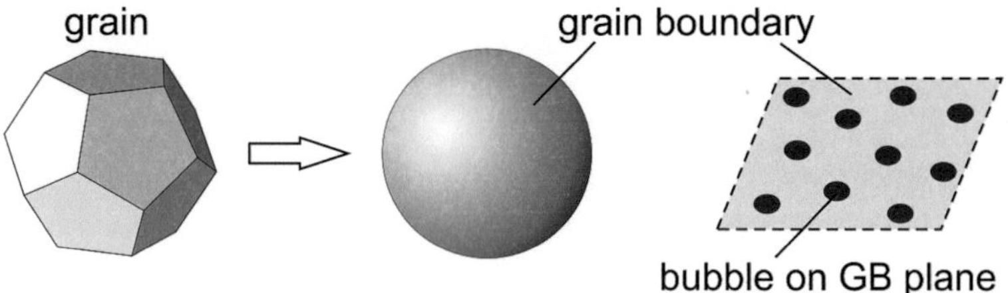

Figure 4.3.: Basic idea of the modified two-dimensional model. Grains are assumed spherical. Helium bubble nucleation and growth happens homogeneously on the whole grain boundary (GB) area.

As an input parameter, time dependent captured helium concentration specific for different material and irradiation conditions is derived from the corresponding 3-D simulations run, integrating Eq. (4.30). 3-D and 2-D simulations are not coupled, meaning that they have to be carried out consecutively. In addition, this prerequisite does not allow emission of helium from GB sinks back into the matrix. The relation of grain surface, A_{gr} to grain volume, V_{gr}

$$\frac{A_{gr}}{V_{gr}} = \frac{3}{R_{gr}}$$

(4.33)

is used to recalculate the helium concentration from atoms per volume to atoms per area, distributing the captured helium atoms throughout the whole GB plane:

$$C_{i,\text{2-D}} = C_{i,\text{3-D}} \frac{R_{gr}}{3}.$$

(4.34)

For the capture rate, Eq. (4.10) has to be changed to

$$A_{i,\text{2-D}} = 2\pi R_i,$$

(4.35)

leading to a capture rate in the 2-D model described by

$$k_{i,\text{2-D}} = 2\pi D_{\text{He}}^{\text{eff,2-D}}.$$

(4.36)

Helium solubility, calculated for the volume by Eq. (4.26), has to be assessed for the GB plane. Still assuming a bcc lattice structure, a 2-D monolayer contains two atoms within a closed packed (110) plane of area $\sqrt{2}a_{\text{Fe}}^2$ down to one atom within a lowest packed (100) plane of area a_{Fe}^2. Since GBs typically are highly unordered regions, the interatomic spacing of iron lattice atoms is quite large, providing room for a large amount of interstitial helium atoms. Simulations are performed starting with a 2-D helium solubility directly re-calculated for a closed packed bcc plane from the 3-D value, leading to

$$C_1^{\text{eq,2-D}} = C_1^{\text{eq,3-D}} \frac{a_{\text{Fe}}}{\sqrt{2}}.$$

(4.37)

Subsequently for a parameter study, the solubility is increased to higher values (for estimating the effects of a one order of magnitude higher solubility in GBs than in the bcc lattice).

Diffusivity of helium in GBs also has to be re-assessed because the unordered GB provides faster diffusion paths due to a lower atomic density. In [108], migration energies E_m for a helium-divacancy cluster in α-Fe were calculated by MD simulations, comparing values for the bulk and two different GB modifications. It was shown that bulk diffusion has a high migration energy of 1.13 eV, while the lowest value of 0.9 eV was calculated for a $\Sigma 3$ GB. For comparing the diffusivities for these two values, a standard diffusion equation of the type

$$D = D_0 \exp\left[-\frac{E_\mathrm{m}}{k_\mathrm{B}T}\right] \qquad (4.38)$$

is used. Assuming a constant pre-exponential factor D_0, calculations for a temperature of 523 and 611 K lead to an enhancement in helium diffusivity by a factor of 160 and 80, respectively, when compared to the bulk value. Since other GB modifications achieve higher migration energies, and the 2-D model presents a mean characteristics approach, the 2-D helium diffusivity is varied between the 3-D and an 80 times larger value.

4.9. Assessment of helium induced hardening

Simulation results from the model developed in this work yield helium bubble size distributions, which are assessed concerning hardening by applying the Dispersed Barrier Hardening (DBH) model. The complete size distribution is taken as an input parameter, and hardening from bubbles of different sizes with corresponding densities is thereby added by using the root sum square (RSS) rule (see Section 3.3). Taking into account bubble sizes ($2R_i$) and densities (C_i) the increase in yield strength is described (analog to Eq. (3.12)) by the equation

$$\Delta\sigma_\mathrm{He} = M_\mathrm{T}\alpha_\mathrm{He}\mu b_\mathrm{disl}\sqrt{\sum_{i=1}^{i_\mathrm{max}} 2R_i C_i}, \qquad (4.39)$$

where the Burgers vector, b_disl of the dislocation for the closest packed $\langle 111 \rangle$ glide direction in the bcc lattice is calculated with

$$b_\mathrm{disl}(T) = 0.5 a_\mathrm{Fe}(T)\sqrt{3}. \qquad (4.40)$$

The barrier strength α_He of helium bubbles is assumed constant for all bubble sizes, and varied between 0.2 and 0.4 within ranges proposed in [72].

5. Numerical code development

5.1. Hard- and software

Calculations are performed on a desktop PC with an Intel® Core™ i7 CPU 860, using four cores with 2.8 GHz and 4 GB RAM. The installed operating system is Windows 7 Enterprise, 64 Bit edition.

For writing the program, FORTRAN77 [109] is used. Temporarily, subroutines from HSL Mathematical Software Library [110] were implemented to verify the numerical solutions of the used diffusion equations on a second path, while the final code does without them. The code is written with Microsoft® Visual Studio® Professional 2008 [111], and it is processed by Intel® Visual Fortran Compiler Professional Edition 11.1 [112]. The code is partially parallelized by using OpenMP® API specification [113] for parallel programming.

5.2. Numerical code

The rate equations of the developed model are numerically solved by a Fortran code. The program structure is shown in Fig. 5.1 and described in the following. At the start, main parameters are chosen, i.e. maximum size of considered bubbles $i_{\max}$, physical constants and also capture and emission rates for all bubble sizes. A quick selection is implemented, to choose between main model parameter values and usage of certain model descriptions, e.g.:

- IRR_EXP causes usage of model parameters characteristic for the different irradiation experiments SPICE and ARBOR1, but also expected fusion conditions, e.g. damage rate and temperature,

- MAT distinguishes between the different irradiated boron doped materials ADS2 and ADS3, e.g. ^{10}B content,

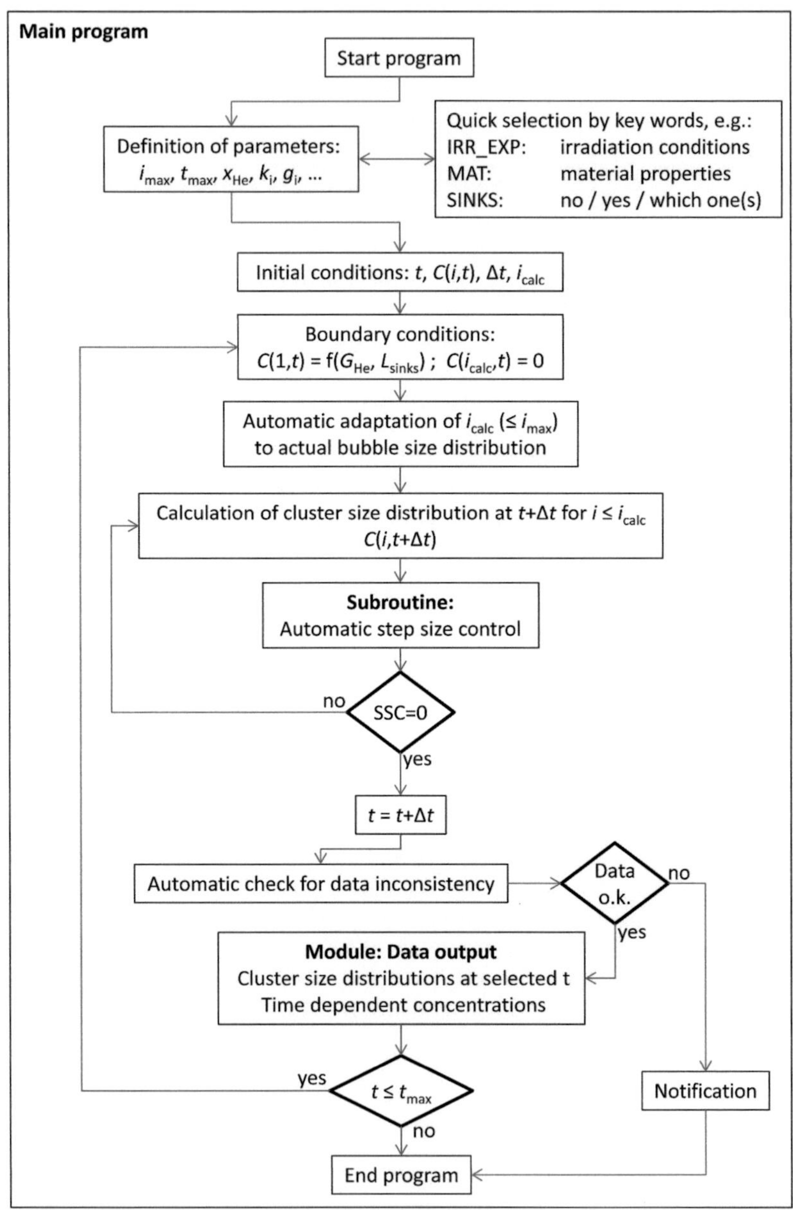

Figure 5.1.: Program structure of developed Fortran code for solving the rate equations.

- SINKS activates helium capturing by chosen sinks.

After defining initial conditions for numerical calculations, e.g. starting time, bubble size distribution, calculation step size, the program enters the calculation and iteration part.

Iteration starts with calculating the actual monomer concentration in the matrix, which serves as left boundary condition for the numerical solving (see Section 5.2.2). Not the whole array of cluster sizes to i_{max} is used for each iteration, but only bubble concentrations to a size of i_{calc} are calculated as described in Section 5.2.2. After calculating $C(i, t + \Delta t)$ subroutine *automatic step size control* is accessed, which adapts the time step as described below in Section 5.2.3 to fit calculation accuracy. After finding an adequate step size, the program leaves the loop and accepts the actual bubble size distribution. A subsequent search for data inconsistency is used to check e.g. for Not a Number (NaN) data and influences of the size of the array holding bubble sizes up to i_{max}. At the end of an iteration, bubble size distribution at specified simulation times and actual concentrations of monomers, cluster content, cluster density and helium concentration at sinks are saved to files. If the simulation time has not reached the target time t_{max}, iteration starts for the next time step, otherwise the program ends.

Further information about the numerical approach is given in the next sections.

5.2.1. Basic numerical scheme

The Master Eq. (4.3) is solved by the forward Euler method [114]. The derivative of the concentration $C(i, t)$ with respect to time is expressed by the differences of the concentrations at $t + \Delta t$ and t for each time step Δt. It must be mentioned that the concentration $C(i, t)$ is denoted slightly different from the previous paragraphs to achieve better clarity. The concentration at time $t + \Delta t$ can be derived from the actual concentration at time t by the following simple scheme.

$$C(i, t + \Delta t) \approx C(i, t) + \Delta t \cdot \frac{\partial C(i, t)}{\partial t} \tag{5.1}$$

These equations are solved numerically by a Fortran code. The required initial and boundary conditions are defined in the next section.

5.2.2. Initial and boundary conditions

Concentrations of different clusters i are assumed zero at $t = t_0$ because there is no helium existent in the matrix at the beginning of the irradiation.

$$C(i,t_0) = 0 \tag{5.2}$$

A one-dimensional matrix $C(1; 2; ...; i_{max})$ specifies the concentrations for all clusters at a given time. With each time step Δt the new values are calculated by a recursive Fortran algorithm using Eq. (5.1).

The left boundary condition is given by the actual helium monomer concentration $C(1,t)$ in the matrix.

$$C(1,t) = \int_0^t G_{He}\, dt - \sum_{i=2}^{i_{max}} iC(i,t) - \int_0^t L_{sinks}\, dt \tag{5.3}$$

It depends on the helium generation rate G_{He} as it will take place in considered experiments or upcoming fusion reactors. G_{He} will be adapted to the experimental conditions to obtain relevant simulation results. The actual helium monomer concentration is decreased by all atoms already bound to clusters. Additionally, helium atoms trapped at sinks cannot take part in clustering any more. The general loss rate to sinks is described by the parameter L_{sinks}.

The right boundary condition is given by the concentration of the largest cluster $C(i_{max},t)$, which is set to zero for all t.

$$C(i_{max},t) = 0 \tag{5.4}$$

That means, the cluster growth must not attain clusters with that size, therefore i_{max} must set large enough. To avoid numerical errors it is important that the cluster density smoothly decreased towards i_{max} at all times t so that no jump occurs.

To avoid large calculation times due to non-allocatable arrays in Fortran77, the code part in Appendix B.1 shows how the right boundary i_{calc} is automatically adapted when the non-zero values of the size distribution approach. For calculations only a small part of the whole array, holding size and corresponding concentration data, up to a bubble size i_{calc} is used, slightly larger than the actual existing largest bubble size. Since the remaining larger bubble sizes have been initiated with a concentration of zero, they are not touched and calculations are not performed for these sizes. Nevertheless, when simulation parameters lead to a large bubble growth, calculation time increases.

5.2.3. Automatic step size control

The calculation time and numerical stability are improved by an automatic step size control algorithm. It is implemented into the Fortran code and adapts the step size Δt in such a way that the change of the concentrations $C(i,t)$ at each calculation step is limited to a maximum value. If the helium bubble growth in the matrix is fast, then the step size is reduced leading to a stable numerical calculation. Whereas if the process approaches a stationary state with little changes of concentrations, the step size and thus the calculation speed are increased.

The used tolerance criterion is provided by [115]. The allowed change of concentration (err) per time step depends on chosen absolute and relative tolerances, tol_{abs} and tol_{rel}, respectively. The maximum function "max [ ]" returns the highest value of the included elements.

$$err \leq tol_{abs} + \max\left[C(i,t), C(i,t+\Delta t)\right] \cdot tol_{rel} \qquad (5.5)$$

In Appendix B.2 the subroutine *stepsizecontrol* is presented which was developed to adapt the step size to calculation properties. Calculation errors are compared by varying the initial step size Δt from the last time step in the following way:

1. If calculation errors for an iteration step using Δt lie within the acceptable error calculated by Eq. (5.5), Δt is doubled until the error gets too large. The largest possible step size within the error range is then taken as calculation parameter for the actual numerical iteration step.

2. If numerical errors for Δt do not fit requirements, step size is divided in half as long as errors are too large. The highest possible time step Δt is then used for the iteration step.

Fig. 5.2 shows a schematic of the program code presented in Appendix B.2.

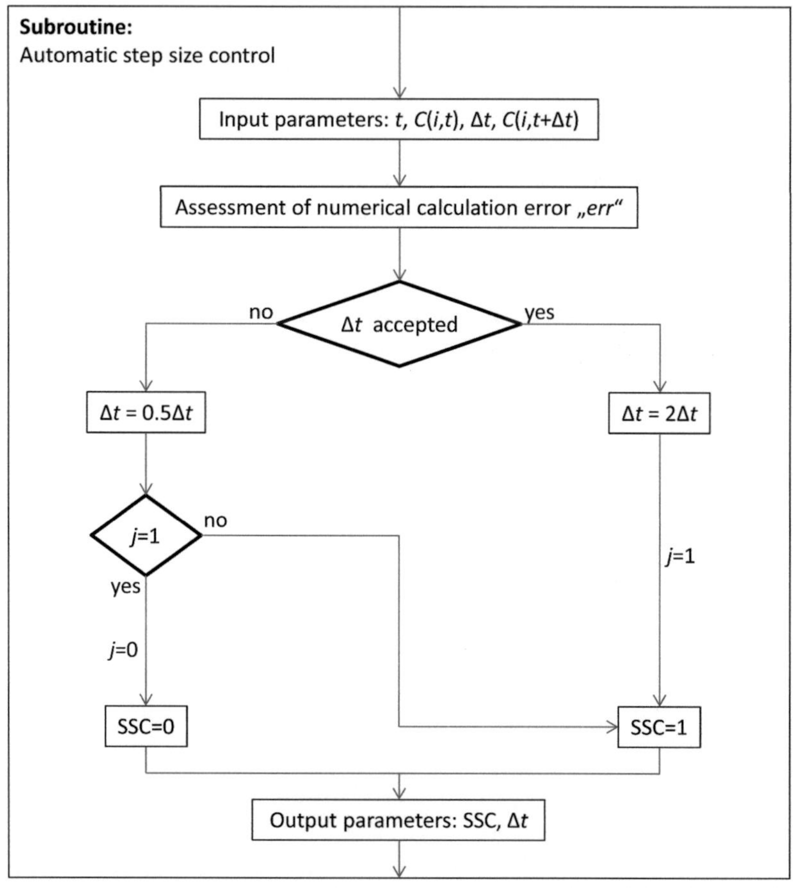

Figure 5.2.: Schematic of subroutine *Automatic step size control* analog to the program code shown in Appendix B.2.

5.2.4. Dealing with numerical instabilities

At several times numerical problems appeared during code development, although the automatic step size adaptation described above was used. Reasons for numerical errors are mainly assigned to accuracy of calculations, in connection with achieving reasonable calculation times. In Fig. 5.3 this problem is exemplarily presented based on prior calculations with simulation parameters close to ADS3 in SPICE, where the effect of the numerical error could be observed very clearly. Fig. 5.3a shows the typical concentration evolution up to irradiation times of 6×10^5 s. After that, the monomer concentration suddenly starts to fluctuate with jumps over several orders of magnitude (straight lines). The reasons for this behavior lie in the accuracy of the calculations, and are produced by the left boundary condition of the numerical solver. In Eq. (5.3), the actual monomer concentration must be calculated for each iteration, and consists of the difference between cumulative generated helium and the sum of helium bound to clusters of sizes two and larger (without considering sinks). Imprecision induced by calculation errors is cumulatively summed up over a large number of calculation steps. When the cluster content increases, and almost all helium is bound to clusters, the difference in Eq. (5.3) becomes very small. Since the overall generated helium concentration is calculated analytically, inaccuracy in the amount of helium bound to clusters may exceed the maximum concentration leading to negative values for the monomer concentration. It was shown that an increase in accuracy by decreasing tol_{rel} in Eq. (5.5) shifts the numerical error to later irradiation times. Under certain simulation conditions, especially decreasing or ceasing helium generation rates, however, a high accuracy could not prevent its occurance. Additionally, calculation times especially for simulations for the SPICE experiment became unacceptably long.

The effect of the error on the cluster size distribution is presented in Fig. 5.3b, where the steep increase in monomer concentration causes a high amount of newly generated small clusters. The blue shaded area shows the amount of the wrongly added helium concentration, which after 10^6 s has already evolved towards larger cluster sizes forming a second peak. This increase in cluster density can also be observed in Fig. 5.3a, where the corresponding curve bends up at the time the error occurs. This simulation aborted after these numerical problems, but it has to be mentioned that some calculations, where this effect was less distinctive, proceeded and produced several other peaks in the bubble size distribution.

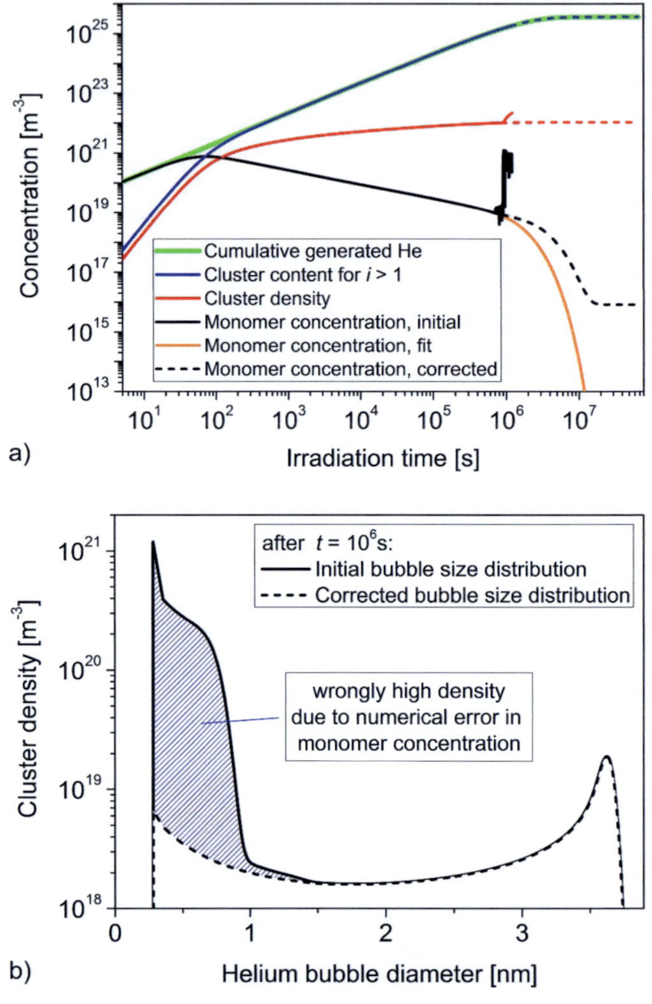

Figure 5.3.: a) Simulation showing effect of numerical instability (straight lines) during calculations. Dashed lines present simulation with fitted and corrected monomer concentration. b) Corresponding bubble size distributions after 10^6 s showing wrongly increased density of small bubbles (blue shaded area) due to numerical calculation error.

To solve this problem, the helium monomer concentration is fitted by an equation showing the typical curve development known from previous calculations (orange curve in Fig. 5.3a), in the form of

$$C(1,t) = A_1 \exp\left(-t/t_1\right) + A_2 \exp\left(-t/t_2\right). \tag{5.6}$$

The fit replaces the calculated helium monomer concentration at $t = t_{\mathrm{corr}}$, shortly before the instability occurs. Since the fit presents only an approximation of the real curve development, it over- or underestimates the real monomer concentration in the matrix. This will change the helium bubble evolution leading to a change in the helium concentration bound to clusters. To keep the concentrations within the requirements, based on the fitting curve the helium monomer concentration is adapted in such a way that the helium concentration bound to clusters stays close to the cumulative generated helium concentration within a specified error range. This is done by using a corrector term shown in Appendix B.3, where starting after t_{corr} the difference between the two concentration is kept within 0.1% of the generated helium concentration, and the monomer concentration is adapted by increasing or decreasing its value by 1% until the requirements are fulfilled. The additional program code takes effect after adapting the step size, and re-enters the code when activated before solving the Master equation. It is executed at every second run to increase calculation speed.

The results of the fitting and correction is shown by the dashed lines in Fig. 5.3. The evolution of the different concentrations proceeds within the specifications, and simulated size distributions show the expected behavior. Although simulations using corrected monomer concentrations take longer calculation times, target simulation times of 6.67×10^7 s characteristic to the SPICE irradiation experiment could be achieved. It has to be mentioned that the usage of the monomer correction was only necessary for a few irradiation conditions (especially simulations for SPICE), and the example presented here showed the highest effect of the numerical error on bubble evolution.

6. Simulation results

6.1. Helium generation rates

Simulations need helium generation rates as input parameters for the calculations, i.e. the left boundary condition, Eq. (5.3), has to be defined. The rates are adapted to characteristic irradiation conditions, may it be specifications from irradiation experiments SPICE and ARBOR, or expected conditions under fusion.

6.1.1. Boron doped steels in SPICE and ARBOR1

Simulations are performed on the boron doped alloys ADS2 and ADS3 irradiated in SPICE and ARBOR1 experiments. The overall concentration of generated helium in the steel matrix is calculated by combining the helium production from boron transmutation and the helium amount that is produced by transmutation of other matrix elements for SPICE [116, 117] and ARBOR [118] irradiation conditions:

$$C^{\mathrm{He}}(t) = C_{10\mathrm{B}}^{\max}\left(1 - \exp\left[-\frac{G_{\mathrm{dpa}} \cdot t}{G_{\mathrm{dpa}}^0}\right]\right) + G_{\mathrm{matrix}}\, G_{\mathrm{dpa}} \cdot t. \tag{6.1}$$

Therein, $C_{10\mathrm{B}}^{\max}$ is the $^{10}\mathrm{B}$ content, G_{dpa} is the damage rate, G_{dpa}^0 is the characteristic damage constant for $^{10}\mathrm{B}$ transmutation, and G_{matrix} is the transmutation rate of other helium producing isotopes in the steels and t the irradiation time.

The values for the different irradiation parameters are shown in Tab. 6.1. Fig. 6.1 shows the cumulative helium concentration calculated by Eq. (6.1) for both boron doped alloys irradiated in SPICE and ARBOR1. The boron transmution in the SPICE experiment already ceases after 10^7 s. Further helium is only generated by transmutation of other elements in the steel matrix. The final helium content

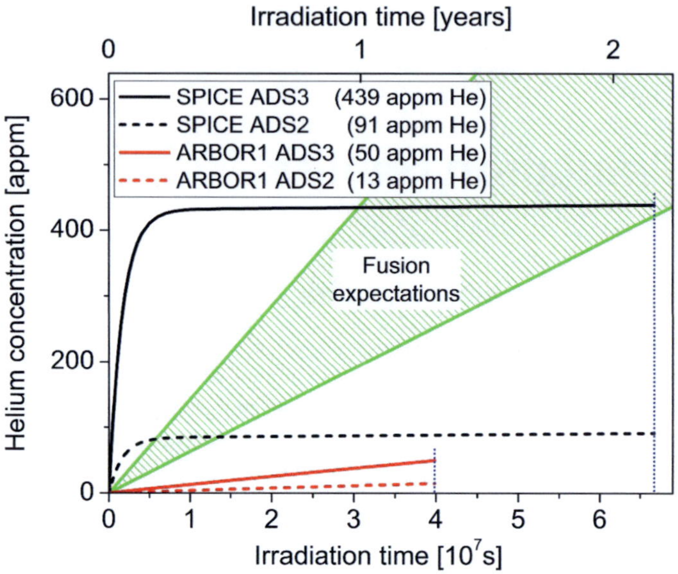

Figure 6.1.: Generated helium in boron doped alloys ADS2 and ADS3 during irradiation in SPICE and ARBOR1 programs calculated by Eq. (6.1). Legend shows values of final helium concentration at the end of irradiation (vertical blue lines). Comparison with expected helium generation rates under fusion conditions is presented.

after 6.67×10^7 s of irradiation in SPICE is about 439 and 91 appm for ADS3 and ADS2, respectively. The situation differs for the boron transmutation in the ARBOR1 experiment: due to the low transmutation cross section of ^{10}B for fast neutrons the generated helium concentration is quite low in comparison to SPICE. Thus the transmutation of steel matrix elements has a higher influence on the cumulative helium content, leading to 50 and 13 appm for ADS3 and ADS2 at the end of irradiation after 3.98×10^7 s, respectively. In the case of ADS2 in ARBOR1 almost the same amounts of helium are produced by ^{10}B and steel matrix elements. Helium generation rates are almost constant up to the end of ARBOR1 irradiation. For simulation purposes, these conditions seem more convenient because

- they are more comparable to fusion conditions (see next paragraph),

Table 6.1.: Parameters of helium generation and irradiation programs. Parameters for the irradiation experiment SPICE are taken from [116, 117], ARBOR1 parameters from [118].

Parameter	Unit	SPICE		ARBOR1	
		ADS2	ADS3	ADS2	ADS3
C_{10B}^{max}	appm	84	432	84	432
G_{dpa}	dpa/s	2.045×10^{-7}		5.620×10^{-7}	
G_{dpa}^0	dpa	0.34		212.30	
G_{matrix}	appm/dpa	0.54		0.29	
Irradiation time t	s	6.67×10^7		3.98×10^7	
Cumulative damage	dpa	13.6		22.4	
Irr. temperature T	K	523.15		611.15	

- coarsening effects like coalescence of bubbles, which are not described by the model, as well as bubble shrinkage due to cascade-bubble interactions will have a greater influence under continuous irradiation if helium production ceases.

6.1.2. Fusion conditions

In fusion power plants, dose rates, G_{dpa}, between 20 and 30 dpa/year with helium generation rates $G_{He/dpa}$ from 10 to 15 appm/dpa are expected in the First Wall (FW) [15]. Transmutation rates are assumed to remain unchanged with time due to a high fraction of helium producing isotopes; e.g. the isotope ^{54}Fe is enclosed at a percentage of 5.8% within iron and produces helium by an (n-α)-reaction. These conditions will result in helium concentrations of 200–450 appm/year.

The actual generated helium amount under fusion conditions is calculated by the equation

$$c^{He}(t) = G_{He/dpa}\, G_{dpa} \cdot t. \tag{6.2}$$

G_{dpa} and $G_{He/dpa}$ are varied within the mentioned expected range, and simulations are performed for some key values.

Table 6.2.: Standard simulation parameters for SPICE and ARBOR1 relevant temperatures with references.

Parameter	Unit	SPICE	ARBOR1	*Ref.*
T	K	523.15	611.15	–
γ_{SF}	J/m^2	2.332	2.318	[100]
a_{Fe}	nm	0.2875	0.2879	[68]
C_1^{eq}	m^{-3}	5.01×10^{-2}	1.61×10^{3}	Eq. (4.26)
D_{He}^{eff}	m^2/s	8.0×10^{-15}	2.4×10^{-14}	Fit, [67]
μ	GPa	77.2	73.2	[119]
b_{disl}	nm	0.2490	0.2493	Eq. (4.40)
$G_{1,sub}^{f}$	eV	3.25		[103]
Z_{He}	–	~ 1		[105]
ρ_{disl}	m^{-2}	1.1×10^{14}		[120]
R_{grain}	µm	10		[5]
M_{T}	–	3.06		[75]

6.2. Results for SPICE and ARBOR1

6.2.1. Nucleation of helium clusters

A matter of particular interest is the nucleation process of helium clusters, i.e. how nucleation and further growth of clusters go hand in hand. Therefore, Fig. 6.2 shows the cumulative generated helium, the helium monomer concentration, cluster content and cluster density for short irradiation times. Calculations are performed for ADS2 and ADS3 in SPICE and ARBOR1 experiments with standard parameters listed in Tab. 6.2 and helium-to-vacancy ratio x_{He} of 1.

Fig. 6.2 shows the same trend for all simulations: the cumulative generated helium concentration (curve 1) is calculated by Eq. (6.1), and partitioned into monomer concentration (curve 2) and cluster content (curve 3). At the beginning of the irradiation, all generated helium is introduced as single substitutional atoms (monomers) into the matrix and contributes to the nucleation phase. At a certain time the monomer concentration undergoes a maximum. At this time the same amount of helium is already bound to clusters, the curves for monomer concentration and cluster content intersect. This indicates that the model follows

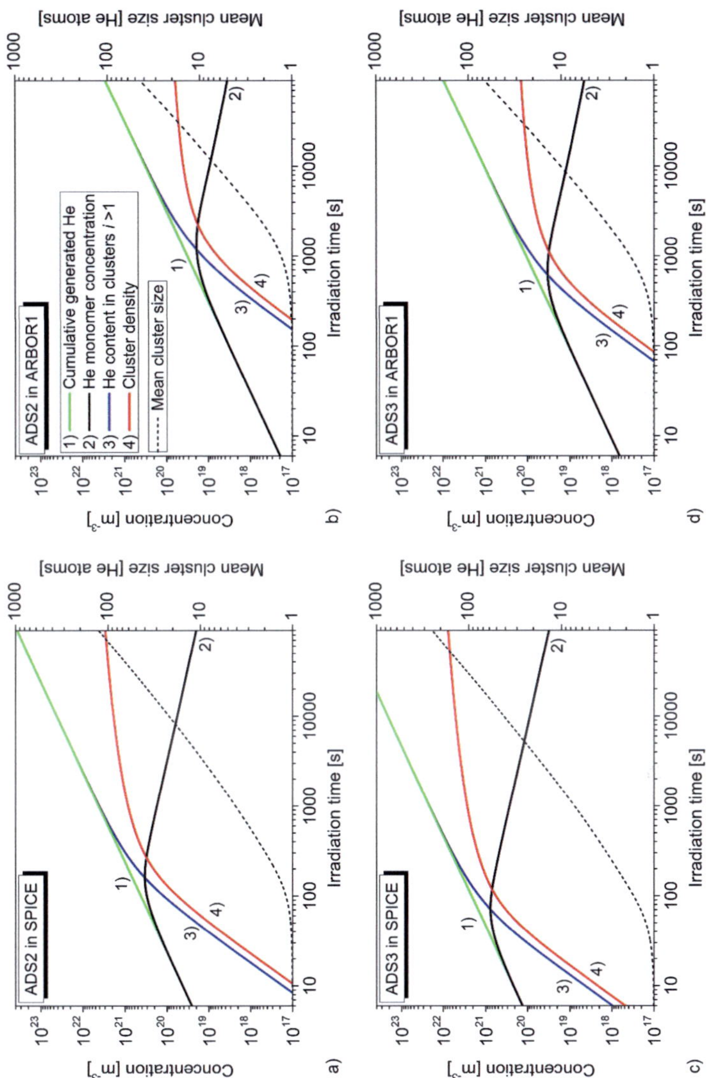

Figure 6.2.: Nucleation phase and further growth of helium clusters in ADS2 and ADS3 at short irradiation times in SPICE (a, c) and ARBOR1 (b, d) experiment. The evolution of generated helium, helium monomer concentration, cluster content and cluster density are shown for $x_{He} = 1$. The mean cluster size (in *number of helium atoms*) is calculated and given by the dashed line. The legend for all four diagrams is presented in b).

the mechanism of di-atomic nucleation, where such behavior is expected [61]. From there on the helium monomer concentration in the matrix decreases, i.e. it is more and more likely that a newly generated helium atom is added to an existing cluster than to find another monomer for di-atomic nucleation. Therefore, the cluster density (curve 4) shifts away from the total helium concentration and starts to saturate.

Differences between the curves in the diagrams in Fig. 6.2 can be observed, which are related to the different irradiation conditions. In fact, two main parameters are responsible for the changes: the irradiation temperature influencing helium solubility and diffusivity, and the helium generation rate due to different ^{10}B doping and neutron spectra. Comparing ADS2 and ADS3 under the same irradiation conditions, one can observe that the peak of the monomer concentration shifts to higher irradiation times when the helium generation rate is lower. On the other hand, a comparison of the same model alloy under SPICE and ARBOR1 irradiation shows that the peak monomer concentration is reached at a different time, which is one order of magnitude higher under ARBOR1 conditions. Although helium generation rates are almost constant within both irradiation experiments in the illustrated time ranges, this effect interferes with temperature influences on simulation parameters like diffusivity and solubility. Since these effects are not separately accessible in this way, a study of all relevant parameters is performed subsequently in Section 6.3.

Using the example of ADS3 in ARBOR1 (Fig. 6.2d), absolute values are discussed next. The helium monomer concentration peaks after 667 s with a concentration of $3.3 \times 10^{19}\,\mathrm{m}^{-3}$. At this time the generated cumulative helium concentration is $7.3 \times 10^{19}\,\mathrm{m}^{-3}$, which corresponds to 0.871×10^{-3} appm. Although slower than in the SPICE experiment, the nucleation phase elapses within a very short time of about 30 minutes compared to the whole irradiation period of more than one year. The curve of the mean diameter (dashed line) of clusters (including clusters with size $i = 1$, i.e. helium monomers) presents a different way of showing these results. While at low irradiation times the mean diameter in terms of *number of helium atoms* is one, it is about 1.46 when the monomer concentration peaks, and reaches 2 after 1200 s of irradiation. The cluster density at the end of the ARBOR1 experiment at $3.98 \times 10^{7}\,\mathrm{s}$ is about $2 \times 10^{20}\,\mathrm{m}^{-3}$, which is only twice the value of the density after 10^{4} s.

Fig. 6.3 shows nucleation rates for all four irradiation cases. Curves were calculated by building the time derivatives of the cluster densities presented in Fig. 6.2.

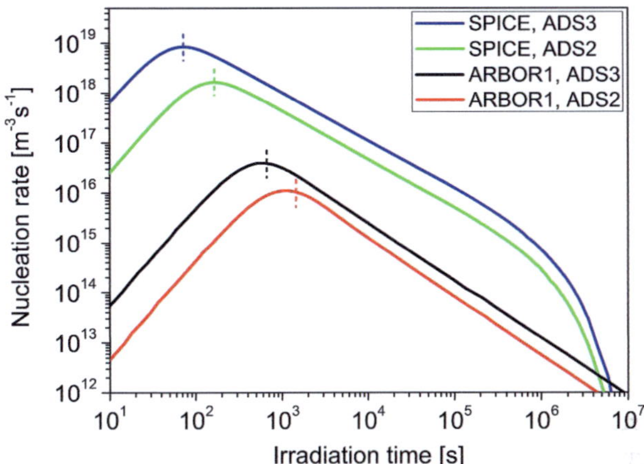

Figure 6.3.: Nucleation rates derived from time derivatives of cluster densities. Dashed lines mark positions of monomer concentration maxima from Fig. 6.2.

Dashed lines show the peak positions of the corresponding monomer concentrations. Both peaks match quite well, although for the experiments in ARBOR1 the nucleation rates peaks shortly before the monomer concentration. Nevertheless, the observed behavior is also an indicator for di-atomic nucleation described in [61]. In SPICE, nucleation rates decrease much faster after 10^6 s, which can be attributed to the boron burn-up and non-linear transmutation rate after this time (see Fig. 6.1).

6.2.2. Evolution of bubble size distributions

The model not only allows description of mean values of the evolving helium bubbles, as they are often presented in literature, but simulations yield time dependent cluster/bubble size distributions (SDs) at every instant of interest.

Fig. 6.4a and b show bubble SDs for ADS3 in SPICE and ARBOR1 experiments, respectively, for bubble sizes larger than 1 nm and characteristic times up to the end of each irradiation experiment. Besides absolute values, ADS2 and ADS3 show the same SD behavior within the same experiment, therefore only a comparison between both irradiation experiments is shown here. After 10^4 s

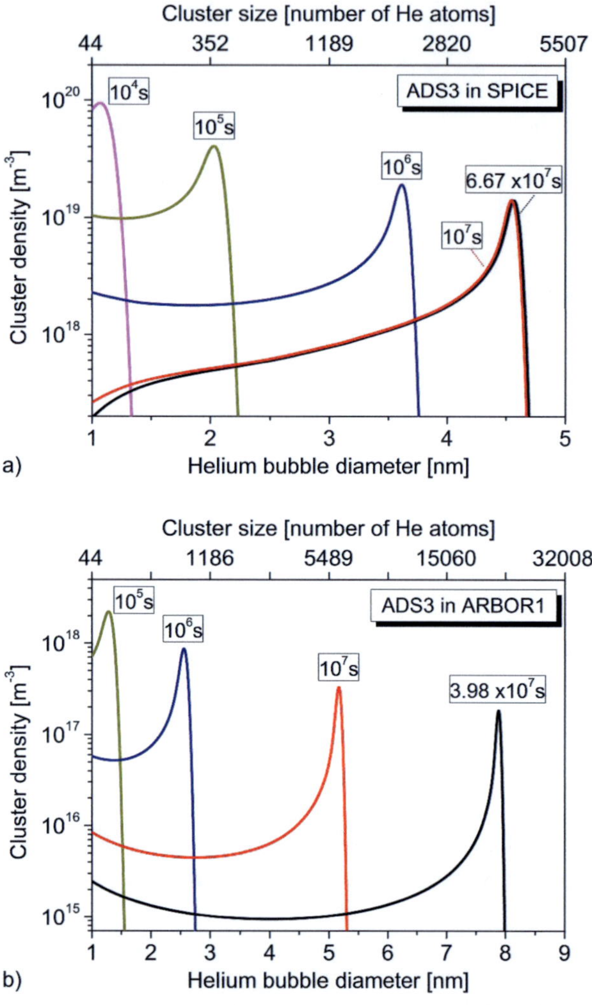

Figure 6.4.: Evolution of cluster size distributions for ADS3 in a) SPICE and b) ARBOR1. Size distributions for characteristic times are presented, also final bubble size distributions at the end of both irradiation experiments. Top x-axis shows corresponding bubble size in terms of *number of helium atoms*. A comparison of the final bubble size distributions for both materials and irradiation experiments is presented in Fig. 6.5.

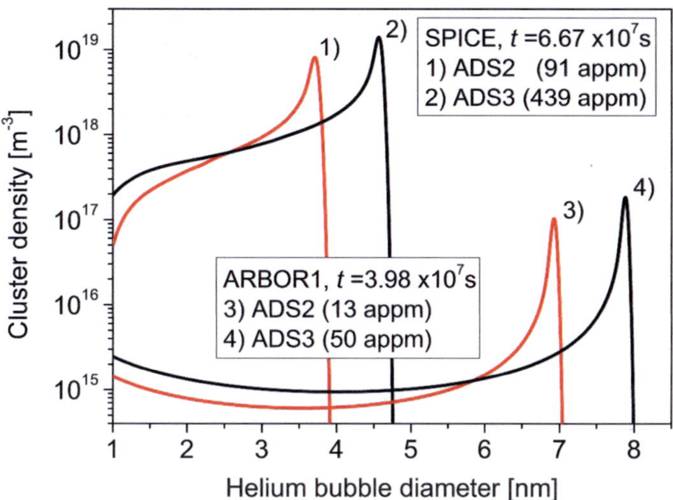

Figure 6.5.: Final bubble size distributions ($x_{He} = 1$) in boron doped ADS2 and ADS3 after SPICE and ARBOR1 irradiation. Legend shows irradiation times and cumulatively generated helium amounts.

bubbles of size of 1 nm are already visible in SPICE, while in ARBOR1 they are still smaller. This is because the incubation phase and nucleation process in ARBOR1 needs a longer time as shown in Fig. 6.2c and d due to a lower helium generation rate and hence a slower helium build-up. Up to 10^5 s peak diameters of ARBOR1 SDs are still smaller than the corresponding SPICE values, but from 10^7 s on peak diameters evolve much faster than in SPICE. Changes in bubble SD from 10^7 s to the end of SPICE irradiation are marginal. The effect is again related to the completed boron burn-up, and only matrix elements produce further helium. This is also the reason for the decrease of concentrations at small bubble diameters, because only a very small amount of helium is produced and introduced into the SD. Helium bubbles tend to coarsen due to the size dependent helium resolution mentioned in Section 4.4.2. In ARBOR1, on the contrary, constant helium transmutation provides new helium at all times, which leads to a high concentration of small helium bubbles throughout the experiment. Peak concentrations decrease with proceeding time, values for SPICE are by more than one order of magnitude higher than for ARBOR1. This is because helium concentration in SPICE is much higher than in ARBOR1, and the area under the curves corresponds to the produced helium amount. Since in SPICE helium diffusivity is lower due to a lower temperature, the shift with time of the peak bubble to larger bubble sizes proceeds slower. A higher pile-up of small and medium sized bubbles is thus produced due to a higher helium generation rate.

Final helium bubble size distributions for both boron doped steels after SPICE and ARBOR1 irradiation are shown in Fig. 6.5. For SPICE, SD peaks at 3.7 nm with a concentration of $8.1 \times 10^{18}\,\mathrm{m^{-3}}$, and at 4.6 nm with a concentration of $1.4 \times 10^{19}\,\mathrm{m^{-3}}$ for ADS2 and ADS3, respectively. For ARBOR1, the corresponding values are 6.9 nm ($1.0 \times 10^{17}\,\mathrm{m^{-3}}$) and 7.9 nm ($1.8 \times 10^{17}\,\mathrm{m^{-3}}$) for ADS2 and ADS3, respectively. Within one irradiation experiment, in ADS2 peak concentrations and diameters are smaller, and also concentrations of small bubbles around 1 nm are less than in ADS3. Overall achieved helium concentrations are given in the legend, and are responsible for the differences between the two materials, together with generation rates and irradiation temperatures.

6.3. Parameter study

It was demonstrated in Section 6.2, that the developed model is capable of providing simulation results with chosen standard parameters for SPICE and ARBOR1 shown in Tab. 6.2. As already discussed in Chapter 4, parameter values hold an uncertainty because a few assumptions and approximations were made influencing the evolution of the bubble size distributions. In this section, an assessment of the parameters in question is provided through a parameter study. It will also be used to prove, that the model is capable of delivering reasonable results besides standard SPICE and ARBOR1 specifications.

Fig. 6.6 shows the complex mutual dependencies of the model input parameters. Arrows demonstrate how the change of one parameter affects the others, and finally influence capture and emission rates of helium monomers. In the case of the helium-to-vacancy ratio, the dashed line shows that the temperature dependency is not implemented in the model. Calculating a size dependent x_{He} by using an appropriate gas equation of state would also lead to a temperature dependence of x_{He}. In the case of the capture rate, a direct exponential temperature dependence is only given when bubble growth is governed by a monomer attachment reaction, otherwise k_i is only affected indirectly through the parameters presented. The

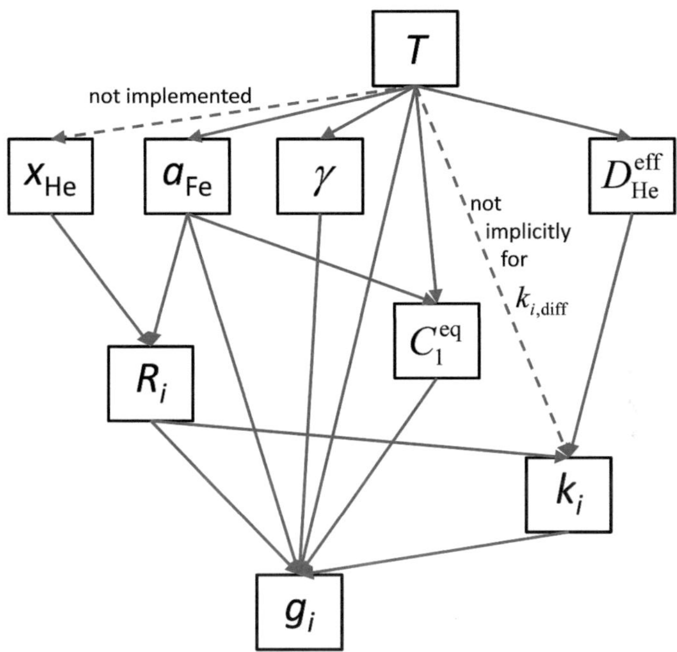

Figure 6.6.: Mutual dependencies of model input parameters. Influences on capture and emission rates are visualized by arrows. Temperature effects on helium-to-vacancy ratio (dashed line) are not described by the model. For the capture rate, direct (exponential) temperature dependence is only given when a monomer attachment reaction barrier is taken into account.

central parameter, which is affected by many variables, is the helium solubility. Since its value is calculated assuming an ideal gas behavior, it could lead to a higher inaccuracy, therefore possible effects on bubble evolution will be evaluated. The lattice constant a_{Fe} is calculated for each temperature using literature data for a Fe-10%Cr steel. This value is close enough to the real EUROFER97 Reduced Activation Ferritic/Martensitic (RAFM) steel matrix, and hence will not be parametrically studied.

In addition to the input parameters shown in Fig. 6.6, helium generation rates are of course responsible for helium bubble evolution. Their effect was shown clearly in Fig. 6.5 when comparing constant but different helium generation rates in ADS2 and ADS3 under ARBOR1 irradiation conditions. Using the basic model configuration, i.e. without helium resolution from bubbles through cascade interaction, the damage rate G_{dpa} only has influence on the helium production, but does not affect other parameters. This is not correct in detail, because neutron irradiation would increase helium diffusivity by Radiation Enhanced Diffusion (RED). This effect is not distinguished for the different irradiation experiments with their specific damage rates, and the same diffusivity, apart from the temperature dependence described, already including irradiation influences is used (see Section 3.2.1).

In the following paragraphs, simulation results with varied parameters described above are shown. Values of remaining parameters are adopted from standard values for ADS3 in ARBOR1 (see Tab. 6.2), if they do not depend on the varied parameter. Calculations are performed for simulation times up to one year, i.e. 3.15×10^7 s, leading to a produced helium amount of 40 appm.

6.3.1. Solubility

Standard solubility under ARBOR1 conditions given in Tab. 6.2 is scaled up and down by one order of magnitude. Fig. 6.7 shows evolution of concentrations and the resulting size distributions after one year of irradiation. It is demonstrated that an increase in helium solubility in the matrix yields a shift in the peak bubble diameter to larger bubble sizes, while a contrary effect is visible for a decreased solubility. The diagram of concentrations' evolution shows that a higher solubility yields a later maximum for the monomer concentration. The incubation and nucleation phase is prolonged because the supersaturation by generated helium atoms is lower. This leads to an overall lower cluster density and therefore larger bubble diameters. Furthermore, a higher solubility increases the helium emission rate from clusters, which itself has a dependence on bubble size during the bubble growth regime favoring larger bubbles to grow and small bubbles to shrink.

6.3.2. Surface energy

Surface energies of iron for various calculations in literature are often assumed to lie between 1 and 2.5 J/m^2 depending on temperature. For standard ARBOR1

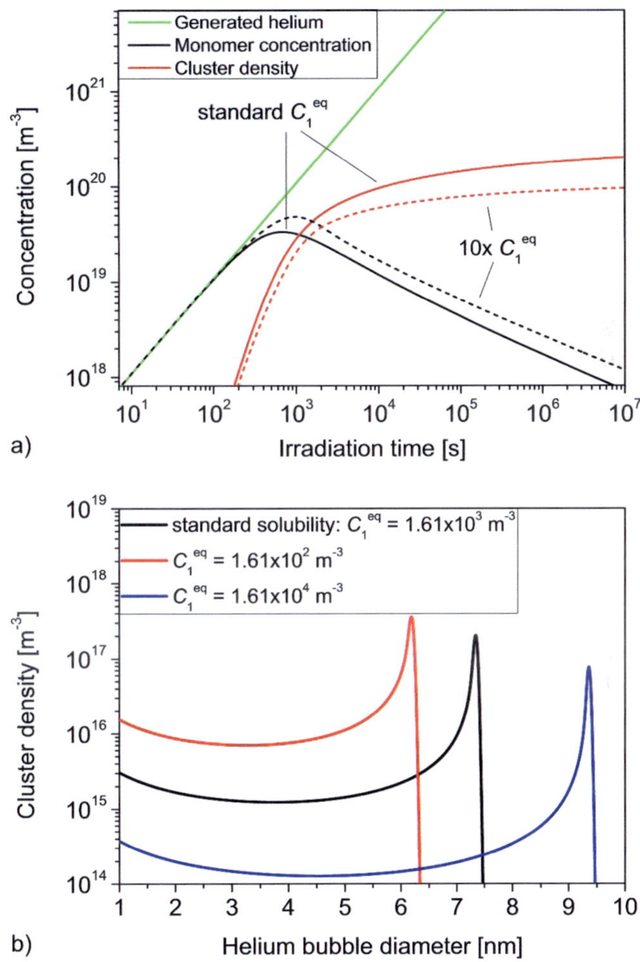

Figure 6.7.: Parameter study of helium solubility C_1^{eq}. Parameter is varied around literature value used as standard simulation parameter. a) Evolution of concentrations, b) bubble size distributions after one year.

conditions, the surface energy used was calculated to be $2.318\,\text{J/m}^2$ based on [100], which lies in the upper region of the literature values. Therefore, the effect of the surface energy on the bubble size distribution is studied by varying γ_{SF} between 1 and $2.5\,\text{J/m}^2$ as shown in Fig. 6.8. Concentration evolution shows the same behavior as was presented when varying the helium solubility: a higher surface energy leads to a lower cluster density due to an altered incubation and nucleation phase. It directly influences the helium emission rate from existing bubbles leading to an increase in peak bubble diameter for higher γ_{SF} when compared to calculations using a lower surface energy.

6.3.3. Helium density in bubble (helium-to-vacancy ratio)

The model allows variation of the helium-to-vacancy ratio x_{He} inside the bubbles. A constant x_{He} is assumed for all bubble sizes, which may not be correct in detail, but is used to understand the influences of different helium densities on helium bubble growth. Fig. 6.9b shows simulated bubble size distributions after $10^6\,\text{s}$ and one year of ARBOR1 irradiation. Since calculations are based on discrete bubble sizes i, curves start at small cluster sizes with different diameters, because clusters with size i have two times (four times) the volume for $x_{He} = 0.5\ (0.25)$ when compared to a helium-to-vacancy ratio of one. After $10^6\,\text{s}$ peak bubble sizes are larger for smaller x_{He}, and diameters differ by 1 nm between $x_{He} = 1$ and 0.25. The remaining amount of helium, since all distributions contain the same helium concentration, is kept by bubbles of smaller size. Their density is much higher for the lower helium-to-vacancy ratios of 0.5 and 0.25. After one year of irradiation, it is observed that the increase of the peak bubble diameter for $x_{He} = 0.5$ is the lowest. The distribution for $x_{He} = 1$ has the same peak bubble diameter, while the distribution for $x_{He} = 0.25$ shows a higher shift to the right. Differences in curves for x_{He} of 1 and 0.5 are mainly observable through a high density of small clusters for $x_{He} = 0.5$. Comparison of $x_{He} = 0.5$ and 0.25 shows that densities of small clusters are similar, and the larger clusters for $x_{He} = 0.25$ contain the remaining helium amount.

Both curves for cluster density in Fig. 6.9a start parallel at small irradiation times, while the values for simulations with $x_{He} = 0.25$ are always higher than for $x_{He} = 1$. After $2000\,\text{s}$ both curves approach quite close but do not intersect, and afterwards the cluster density for the lower helium-to-vacancy ratio increases much faster.

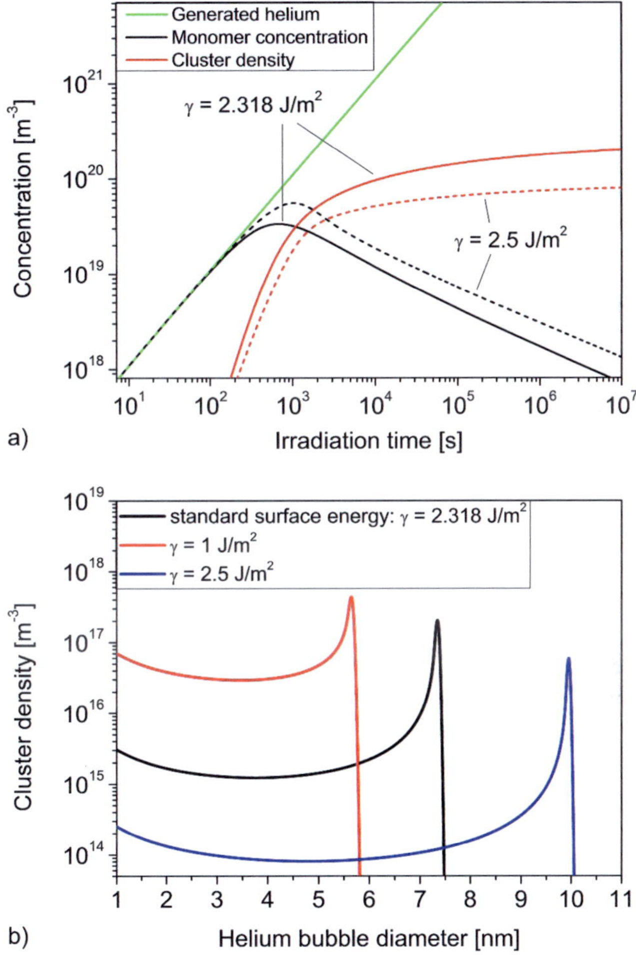

Figure 6.8.: Parameter study of surface energy γ_{SF}. Parameter is varied around literature value used as standard simulation parameter. a) Evolution of concentrations, b) bubble size distributions after one year.

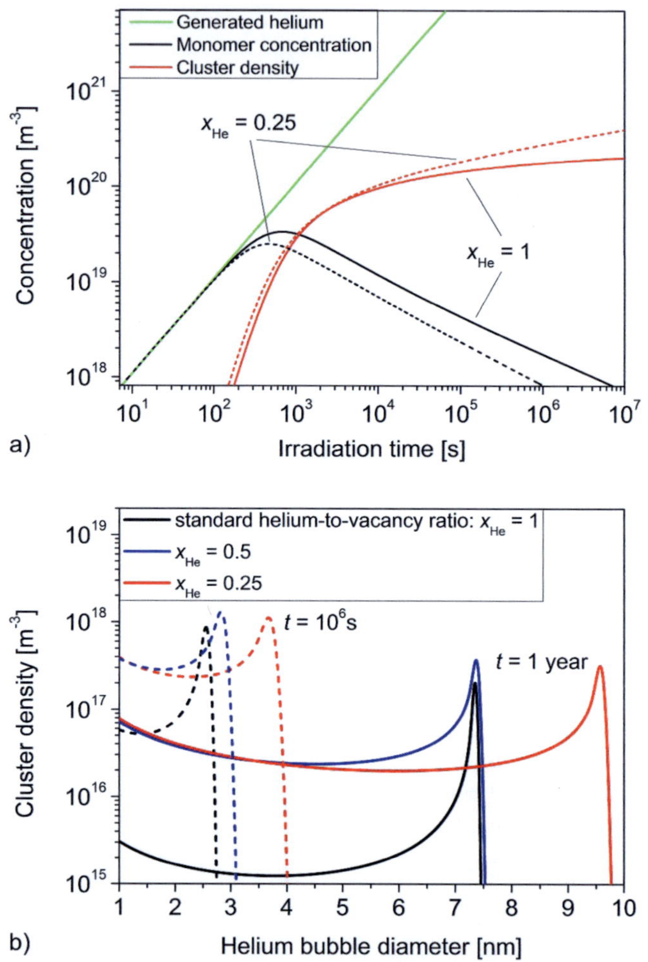

Figure 6.9.: Parameter study of helium-to-vacancy ratio x_{He}. Parameter is varied to 0.5 and 0.25 and compared to standard calculations with $x_{He} = 1$. a) Evolution of concentrations for highest and lowest x_{He}, b) bubble size distributions after 10^6 s and one year.

6.3.4. Diffusivity

Simulations with varied helium diffusivity are shown in Fig. 6.10. Final bubble size distribution after one year is shifted to larger bubble sizes when a ten times higher diffusion coefficient is used as simulation parameter. While simulations with standard parameters yield peak bubble sizes of 7.3 nm, ten times lower and higher diffusivity values cause peak diameters of 4.7 and 11.9 nm, respectively. Higher diffusivities lead to an earlier occuring nucleation phase. In this case the cluster nucleation rate and hence the cluster density is higher at small irradiation times. Though similar in shape, the curve peak of the monomer concentration, which can be directly correlated to the nucleation rate peak as shown in Fig. 6.3, extends over an expanded time interval (note the logarithmic scale in Fig. 6.10a) for the standard simulation. Enhanced nucleation finally leads to a higher cluster density and thus smaller peak bubble diameters for a lower helium diffusivity.

6.3.5. Diffusion and reaction governed capture rate

The standard model uses a capture rate based on a diffusion governed growth process. As described in Section 4.4.1 a combination of diffusion and reaction governed mechanisms is possible, and simulation results will be compared to standard calculations in the following part. The attachment barrier of helium atoms to clusters is assumed to be small, for the parameter study values of 0.06 and 0.17 eV are estimated near to the value for the interstitial helium migration energy. Fig. 6.11 presents simulation results after one year of irradiation. It can be observed that a higher attachment barrier leads to smaller peak bubble diameters due to a reduced monomer capture rate. The peak is broadened towards smaller bubble sizes, which is especially striking for $E_i = 0.17$ eV. Incubation and nucleation phases take place at later times when taking into account an attachment barrier, and cluster density is higher at the end of the simulation. Since there is no general consent in literature about the existence of an attachment barrier for helium atoms to bubbles, not to mention the value for that energy, it can be summarized that a large effect on bubble size distribution can be achieved by applying very low attachment barriers E_i.

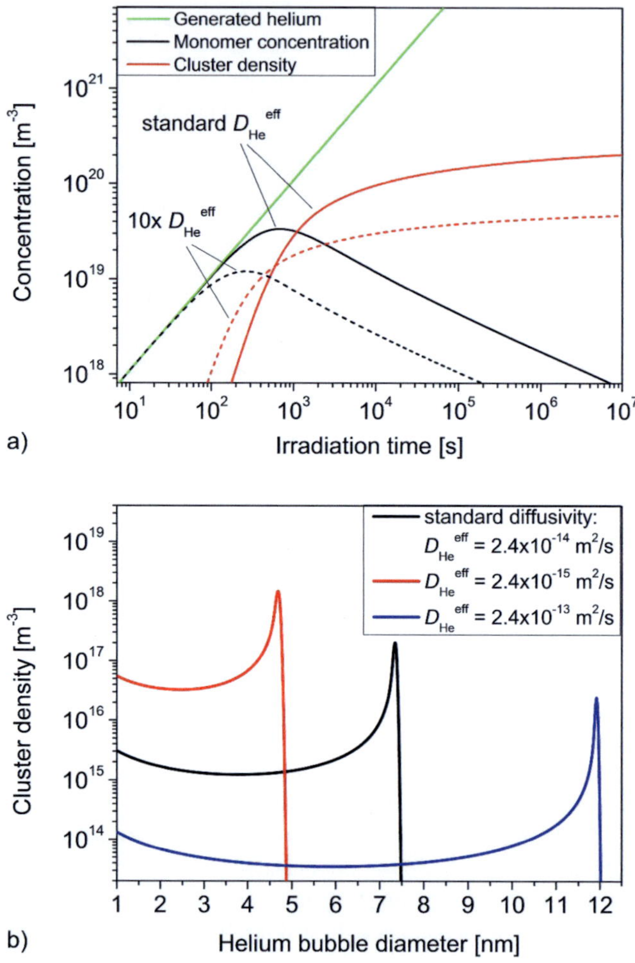

Figure 6.10.: Parameter study of effective helium diffusivity D_{He}^{eff}. Parameter is varied around literature value used as standard simulation parameter. a) Evolution of concentrations, b) bubble size distributions after one year.

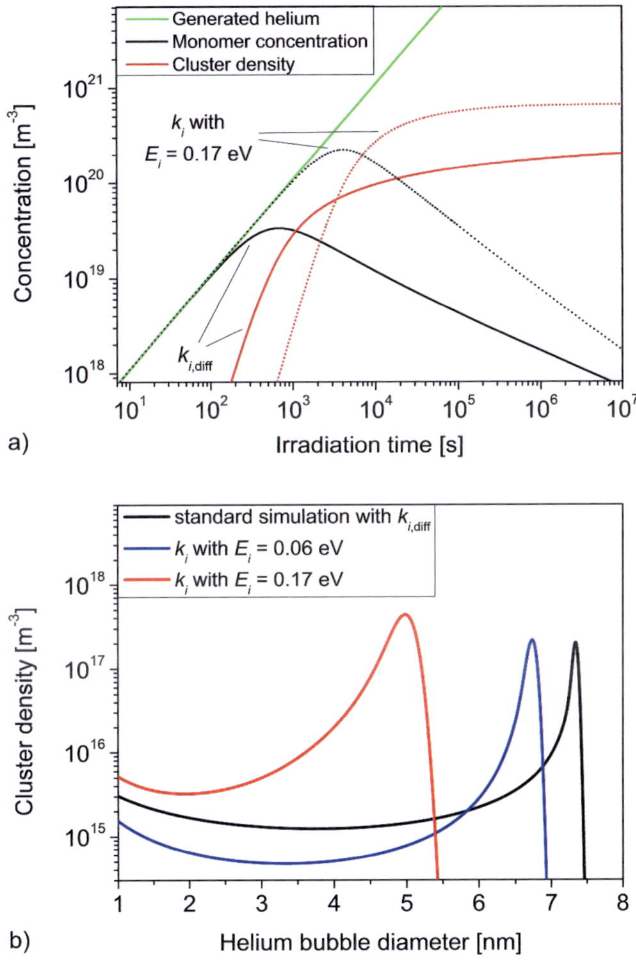

Figure 6.11.: Parameter study when taking into account an attachment barrier E_i of 0.06 and 0.17 cV, yielding a combination of a diffusion and reaction governed capture rate, k_i in accordance to Eq. (4.14). a) Evolution of concentrations, b) bubble size distributions after one year.

6.3.6. Temperature

The influence of the temperature on simulations is presented in Fig. 6.12. A temperature variation alters all important simulation parameters shown in Fig. 6.6, corresponding values are given in Tab. 6.3. A higher temperature has a mixed effect on parameters: while the surface energy is lowered, diffusivity increases by one order of magnitude, and helium solubility even shows a much higher dependance with an increase of three orders of magnitude. The resulting bubble size distribution for $T = 711$ K peaks at 16.5 nm after one year, which is more than twice the diameter when compared to standard simulation conditions. Evolution of monomer concentrations differs significantly from former simulations: up to the maximum curve behaves as seen before, but afterwards a sharp decrease is observed for the high simulation temperature. Towards larger times, monomer concentration shows a recovery and the curve approaches the standard curve again. Also the cluster density shows a lower slope after 1000 s, and final cluster density after 10^7 s is by one order of magnitude lower than for the standard simulation.

6.3.7. Summary of parametric study

Results from parametric study are best summarized by comparing nucleation rates (i.e. time derivative of cluster density) for the different simulation conditions in Fig. 6.13. Black curve describes nucleation rate derived for ADS3 in ARBOR1 with standard parameters. It has to be emphasized that although most curves show similar appearance and supposed peak width, due to a double logarithmic scale of axis the time periods of a corresponding high cluster nucleation rate differ widely. Generally, peak nucleation rates at shorter times are responsible for large

Table 6.3.: Change in temperature dependent parameters caused by varying T in simulations.

Simulation parameter	$T = 511$ K	$T = 611$ K	$T = 711$ K
a_{Fe} (nm)	0.2875	0.2879	0.2882
γ_{SF} (J/m^2)	2.334	2.318	2.302
C_1^{eq} (m^{-3})	9.2×10^{-3}	1.6×10^3	9.4×10^6
D_{He}^{eff} (m^2/s)	2.1×10^{-15}	2.4×10^{-14}	1.4×10^{-13}

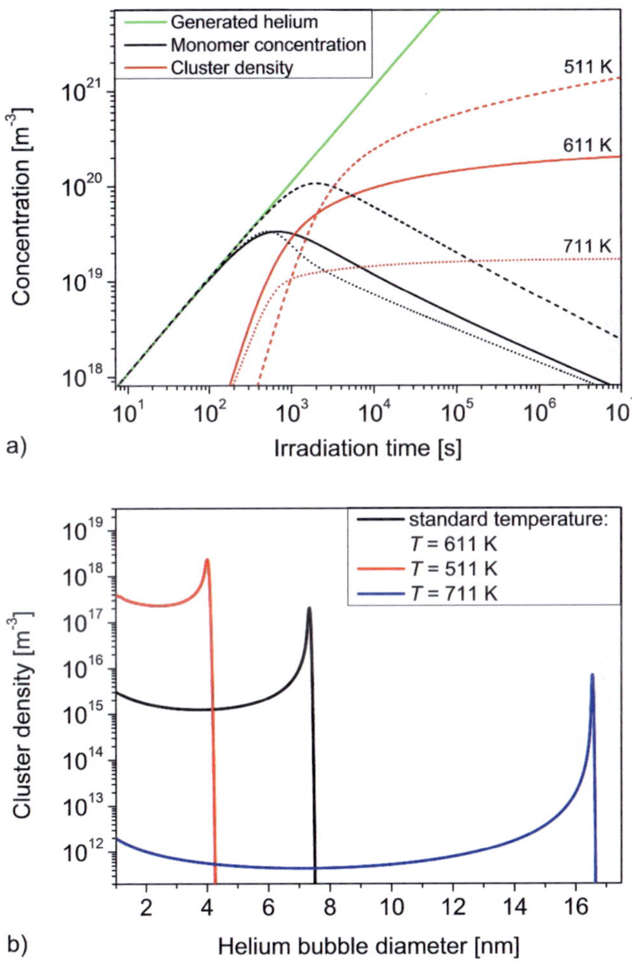

Figure 6.12.: Parameter study when varying temperature T by $100\,\mathrm{K}$, which consequently alters all dependent parameters as described in Fig. 6.6. a) Evolution of concentrations, b) bubble size distribution after one year.

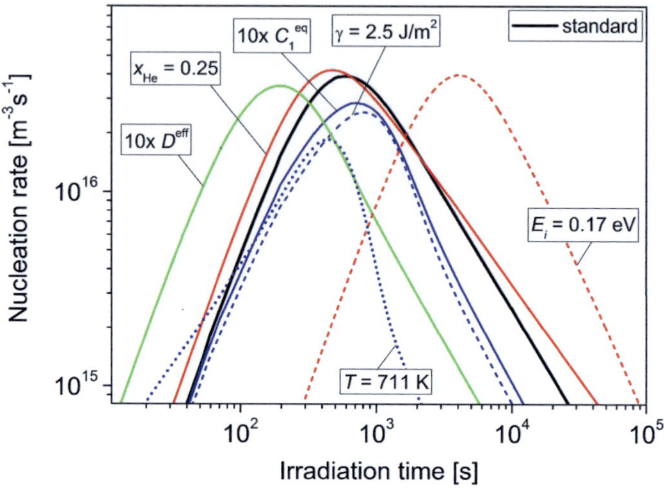

Figure 6.13.: Nucleation rates for different simulation parameters analyzed during parametric study.

peak bubble diameters, while a peak shift of the nucleation rate towards larger times causes a high density of smaller sized bubbles. Simulation parameters that directly influence the emission rate g_i without affecting the capture rate k_i, as shown in Fig. 6.6, cause a change of the nucleation rate's curve behavior. The blue curves for higher solubility, surface energy and temperature show a more and more asymmetric appearance with minor peak width and decreasing peak nucleation rate. Especially towards larger times, nucleation rate decreases become extremely steep with a large difference from the standard curve. The highest effect can be observed at simulations for varied T, which consecutively increased solubility and diffusivity as well.

The parametric study proved that the model is capable of describing helium bubble nucleation and growth over a wide range of simulation parameters. Effects of different parameters on cluster evolution were analyzed. Nucleation rate was presented to be a good indicator to compare different simulation conditions and predict simulation results qualitatively, such as e.g. positions of peak bubble diameters.

6.4. Sink effects

Grain boundaries (GBs) and dislocations (DLs) are considered as sinks for helium, while for the latter case a bias factor Z_{He} of 1 is assumed [105]. With a grain size of 20 μm [5] and a dislocation density of $1.1 \times 10^{14}\,\mathrm{m}^{-2}$ measured by TEM [120] for unirradiated ADS3 the influences of sinks on helium bubble nucleation and final size distribution are determined.

Evolution of relevant concentrations are shown in Figs. A.3 and 6.14 for ADS2 in SPICE (a) and ARBOR1 (b). While absolute concentrations are given in Fig. A.3, Fig. 6.14 shows relative concentrations normalized to the helium concentration generated so far (also compare to [121]). For a chosen irradiation time the fractional helium concentration at different locations can be determined. When considering sinks for helium, the evolution of concentrations shows a different behavior than presented in Fig. 6.2. For the SPICE experiment, the development of monomers, cluster content and cluster density is only slightly influenced by the amount of helium captured at sinks (Fig. 6.14a), because only a small fraction of helium is trapped at both GB and DL sinks. The relative concentration of helium at sinks peaks after 372 s with a total fraction of 8%, and decreases down to 0.3% and an absolute amount of 0.3 appm at the end of irradiation. For ADS2 in ARBOR1, sinks have a great influence on the evolution of concentrations. After about 7500 s, 86% of the generated helium is captured by both sink types (Fig. 6.14b) leading to a decrease in cluster nucleation and bubble growth rates (Fig. A.3b). When the helium monomer concentration in the matrix decreases again, the additional capture of helium by sinks starts to saturate due to the dependence of the sink loss rate on the monomer concentration in Eq. (4.30). Higher temperatures in ARBOR1 increase the fraction of helium at sinks due to the higher diffusivity, yielding a total helium fraction at sinks of 7.2% (equivalent

Table 6.4.: Cumulative helium concentration captured by grain boundary and dislocation sinks at the end of the irradiation experiments. The percentage of the total generated helium in each experiment is given in parentheses. In each case, 92.5% of the listed concentration is situated at grain boundaries, 7.5% at dislocations.

	SPICE	ARBOR1
ADS2	0.3 appm (0.3%)	1.0 appm (7.2%)
ADS3	0.4 appm (0.1%)	1.4 appm (2.7%)

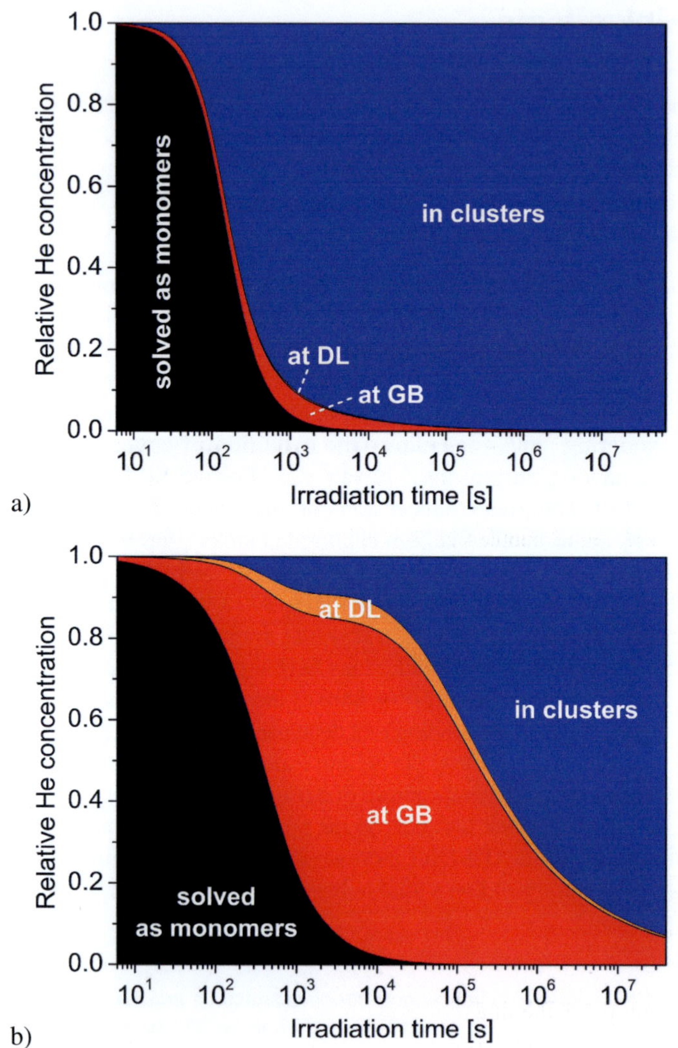

Figure 6.14.: Relative helium concentration normalized to the helium concentration generated so far. For a chosen irradiation time the helium fraction located at the different sites (DL – dislocation, GB – grain boundary) can be determined. Simulations were performed for ADS2 under a) SPICE and b) ARBOR1 irradiation for $x_{He} = 1$. Absolute concentrations are shown in Fig. A.3.

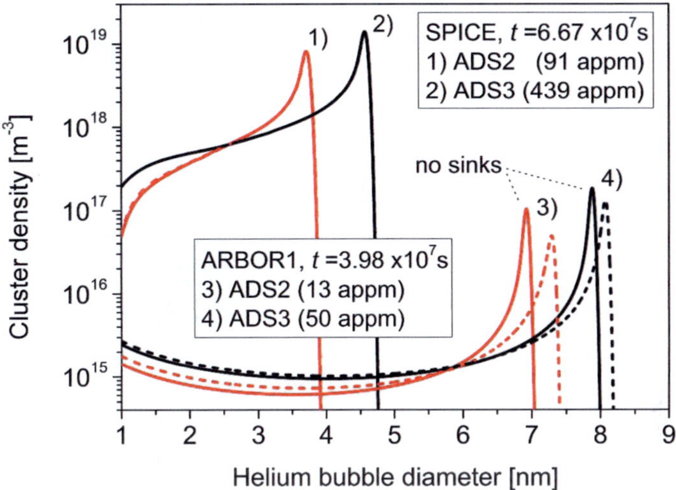

Figure 6.15.: Final bubble size distributions ($x_{He} = 1$) in boron doped ADS2 and ADS3 after SPICE and ARBOR1 irradiation considering grain boundaries and line dislocations as sinks for helium (dashed lines). For the SPICE specimens sinks have no influence on the final bubble size distribution, hence the curves are congruent. Legend shows irradiation times and cumulatively generated helium amounts.

to 1 appm) at the end of irradiation. A summary of the sink effects including simulation results for ADS3 is given in Tab. 6.4. For both irradiation experiments GBs show a higher sink strength than DLs based on the used simulation parameters measured by mentioned TEM investigations on ADS3. The captured helium concentration by sinks for the combinations of both materials and both irradiation experiments is thereby split with a fixed ratio: finally 92.5% ends up at GBs and 7.5% at DLs.

Fig. 6.15 compares the final bubble size distributions resulting from simulations with and without consideration of GB and DL sinks for a helium-to-vacancy ratio of 1. For the SPICE specimens sinks show no influence on size distributions, the calculations yield congruent results. For the ARBOR1 specimens, however, final bubble size distributions change (dashed lines). When taking into account sinks, size distributions are broadened and peaks are shifted towards larger bubble diameters with lower cluster densities. A larger peak shift is observed for ADS2, where the helium generation rate and helium content are lower. Final peak bubble diameters are 7.3 and 8.1 nm for ADS2 and ADS3, respectively.

This effect is attributed to the change in nucleation process, where the concentration of monomers is reduced by the amount of helium trapped at sinks. Therefore the number of nucleation centers decreases leading to lower cluster densities and larger diameters.

6.5. Cascade induced helium resolution from bubbles

Based on simulations for ADS3 under ARBOR1 irradiation, the effect of cascade induced helium resolution from bubbles is investigated. As mentioned in Section 4.7, two different simulation parameters are used for the modification. In case 1, a constant size reduction of 5% is assumed for all cluster sizes, while in case 2 an exponential fit was applied to the Molecular Dynamics (MD) simulation data of resolved helium atoms shown in Fig. 3.14b. Simulation results of both cases for a correlation factor $X_{\mathrm{corr}} = 50$ are presented in Fig. 6.16; for comparison standard simulation without helium resolution is given. The evolution of concentrations is identical for the whole nucleation regime. Afterwards, when cluster growth is dominant, monomer concentration and cluster density increase when compared to the standard curve. The higher increase is observed for the constant cluster size reduction of 5%, while curves for simulations of case 2 lie in between. The cluster density for case 1 after 2×10^7 s is increased by one order of magnitude when compared to standard simulations, for case 2 it is twice the value.

The time evolution of bubble size distributions shown in Fig. 6.16b widely differ when compared to the results of standard simulations without considering helium resolution. In the case for a constant size reduction of 5%, the shift of the peak bubble diameter towards larger bubble sizes with irradiation time is reduced, and even a stationary peak diameter is observed already after 10^7 s. Later on, the size distribution after 2×10^7 s only shows an increase in concentration with slight peak broadening, but the peak diameter remains at 2.9 nm. For case 2, the bubble size dependent helium resolution, a similar behavior can be observed. However, the effect of a decreasing peak diameter shift only starts to appear after

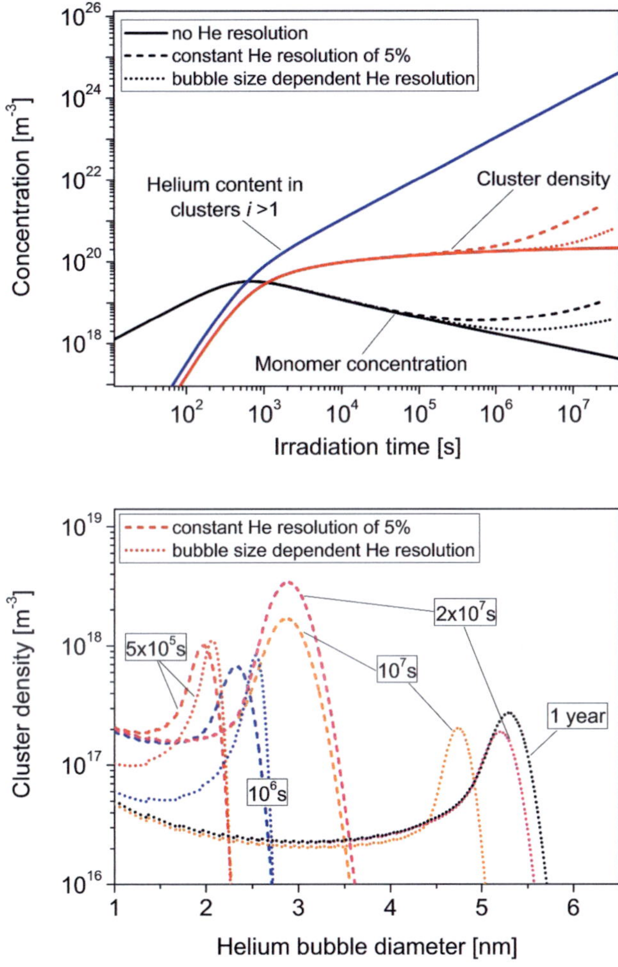

Figure 6.16.: Cascade induced helium resolution from bubbles for a given constant bubble size reduction of 5% (dashed lines) and a bubble size dependent size reduction (dotted lines) based on Fig. 3.14b. a) Concentration evolution, b) time dependent bubble size distributions.

one year of irradiation, and a stationary peak bubble diameter is not yet reached. Nevertheless, the effect of helium resolution from bubbles leads to smaller bubble sizes peaking at 5.3 nm when compared to the peak bubble diameter of 7.9 nm observed after standard simulations for ADS3 in ARBOR1.

Curves of bubble size distribution in Fig. 6.16b show sawtooth shaped lines. This effect stems from the calculation of the size reduction and the following concentration adjustment. Since bubble sizes are given in discrete values of helium atoms, the calculation of a size reduction of e.g. 5% also has to result in integer values. Numerics thereby cut off the calculated modulus of the reduced target bubble size, which is why certain affected helium concentrations from larger bubbles are added to the same smaller bubble during the resolution process.

6.6. Helium bubble nucleation and growth at sinks

By using the modified two-dimensional (2-D) model presented in Section 4.8, simulations are performed for bubble nucleation and growth at GBs using input data from three-dimensional (3-D) simulations on ADS3 in ARBOR1 irradiation.

Helium diffusivity was recalculated to $D_{\text{He}}^{\text{eff,2-D}} = 1.92 \times 10^{-12} \, \text{m}^2/\text{s}$, and helium solubility to $C_1^{\text{eq,2-D}} = 3.28 \times 10^{-7} \, \text{m}^{-2}$. A grain radius, R_{gr} of 10 μm was used. The helium concentration caught by sinks derived from the corresponding 3-D simulation is presented in Fig. 6.17a. Because a continuous function is needed due to different calculation time steps in 3-D and 2-D, in a first attempt a linear curve evolution between data points was approximated. Unfortunately, numerical errors appeared, and the need for a continuously differential function became obvious. Therefore, data was fitted by an exponential equation showing good agreement for large irradiation times. For times less than 10^4 s, however, congruence is bad. Since this time period of less than three hours is very small compared to the whole irradiation experiment, the error in concentration between both curves is less than 0.02%, and hence the fit is used for this simulation.

Fig. 6.17b presents simulated bubble size distributions at GBs sinks. At the end of ARBOR1, a peak diameter of 10.4 nm with a corresponding density of $8.5 \times 10^9 \, \text{m}^{-2}$ is observed. Densities of smaller bubbles are by 4 orders of magnitude lower than the peak density. The peak diameter is by 2 nm larger than for the 3-D simulation.

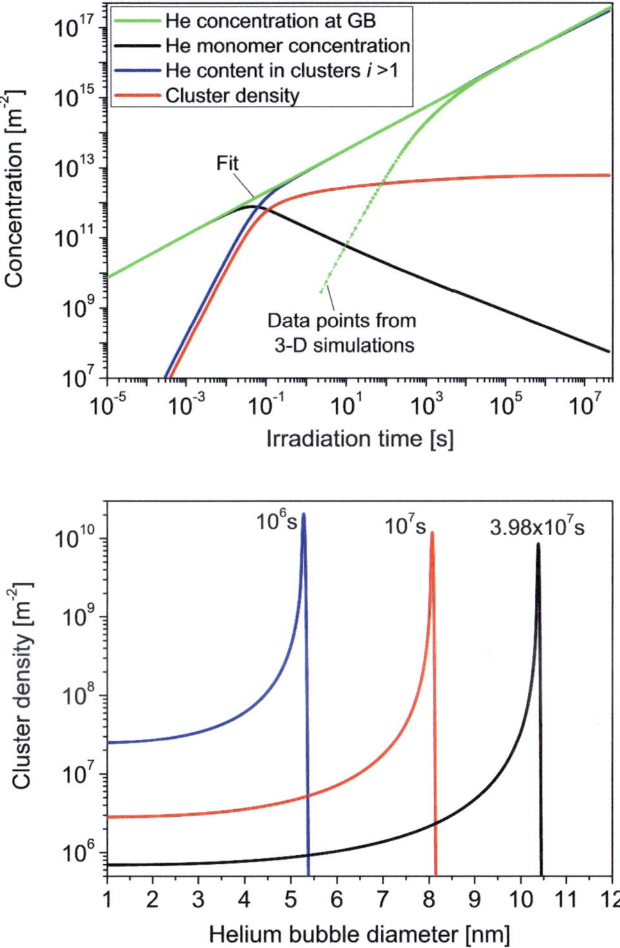

Figure 6.17.: 2-D simulation for helium bubble nucleation and growth at grain boundary sinks. 3-D simulation of ADS3 in ARBOR1 provides input data. a) Concentration evolution, captured helium generation by GB sinks is taken as input parameter and is fitted by a continuously differentiable function. b) Size distributions for different irradiation times up to the end of the ARBOR1 experiment.

6.7. Helium induced hardening

The Dispersed Barrier Hardening (DBH) model is applied to the simulated helium bubble size distributions shown in Fig. 6.5. Taking into account bubble sizes ($2R_i$) and densities (C_i) the increase in yield strength is described by the equation

$$\Delta\sigma_{He} = M_T\alpha\mu b_{disl}\sqrt{\sum_{i=1}^{i_{max}} 2R_i C_i}\,. \qquad (6.3)$$

In this case, also a root sum square (RSS) rule is used to sum up the effects of all different bubble sizes. Parameters are taken from [119, 75], while the barrier strength α for helium bubbles is taken as a fit parameter in the range between 0.2 and 0.4, which classifies helium bubbles as weak/intermediate obstacles to dislocation glide. Tab. 6.5 shows results of simulated hardening due to helium bubbles for both boron doped steels. For ARBOR1, hardening was additionally calculated taking into account effects of sinks.

Hardening calculated with helium bubble size distributions for the ARBOR1 experiment yielded 9–18 and 14–28 MPa for ADS2 and ADS3, respectively. Hardening results slightly decrease when GB and DL sinks are taken into account due to the change in cluster size distributions (Fig. 6.15). For a helium-to-vacancy ratio of 0.5, simulations for ARBOR1 show an increased expected hardening between 14–28 MPa for ADS2 and 21–43 MPa for ADS3. In SPICE, a relatively

Table 6.5.: Helium induced hardening $\Delta\sigma_{He}$ using simulated size distributions from SPICE and ARBOR1 irradiation ($x_{He} = 1$, if not mentioned otherwise). Differences in hardening when taking into account sinks and varying the helium-to-vacancy ratio are shown. The range of the hardening values stems from varying the barrier strength α from 0.2 to 0.4.

		$\Delta\sigma_{He}$ [MPa] in	
		ADS2	ADS3
SPICE	standard	44 - 89	79 - 159
	$x_{He} = 0.5$	49 - 97	86 - 173
ARBOR1	standard	9 - 18	14 - 28
	+ sinks	8 - 16	14 - 27
	$x_{He} = 0.5$	14 - 28	21 - 43

small change of hardening for $x_{He} = 0.5$ is observed. It has to be emphasized that results of hardening calculations decribe only the effect for helium being the exclusively present obstacle to dislocation glide, i.e. the interaction and superposition with other defect types and their hardening contribution has to be considered when comparing calculations with experimental results (see Section 7.3).

6.8. Helium effects under fusion relevant helium generation rates

Expected helium generation rates under fusion conditions are given in [39], varying between 200 and 450 appm per year. Simulations are performed here for constant helium generation rates at a temperature of 523 K for $x_{He} = 1$ assessing the influence on bubble size distributions. Simulation parameters at 523 K can be

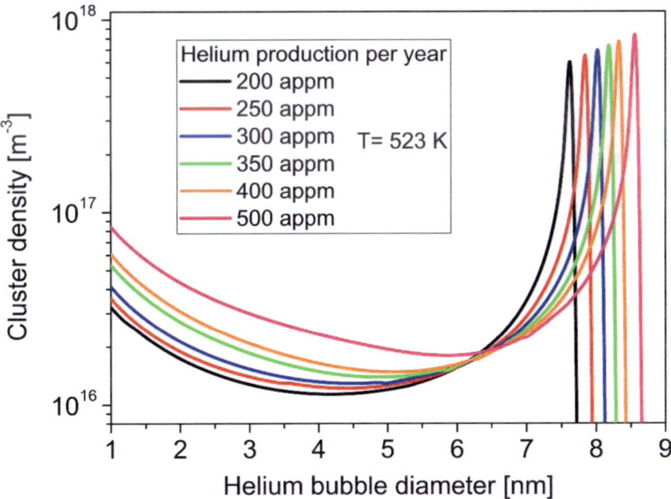

Figure 6.18.: Bubble size distributions calculated for fusion relevant helium generation rates after one year. For a fixed irradiation temperature of 523 K the helium production rate is varied between expected values mentioned in [39].

read out in Tab. 6.2, except helium solubility in the present calculation is taken as $C_1^{\text{eq}} = 0.301\,\text{m}^{-3}$, helium diffusivity as $D_{\text{He}}^{\text{eff}} = 2.4 \times 10^{-14}\,\text{m}^2/\text{s}$.

Fig. 6.18 shows simulation results. A higher helium generation rate leads to larger peak bubble diameters, as well as to higher densities of smaller bubbles. Increasing a helium generation rate from 200 appm per year to twice the value thereby shifts the peak bubble diameter from 7.6 nm to 8.3 nm, and the density of one nanometer sized bubbles increases by a factor of two. Up to sizes of 6 nm, the doubled helium generation rate yields significantly higher bubble densities.

The effect of the density increase should be more noticable in corresponding microstructural Transmission Electron Microscopy (TEM) images than the increase of the peak bubble diameter. It has to be emphasized that the temperature of 523 K presented in these simulations is too low for expected irradiation conditions for structural RAFM steels in the fusion environment, where an operation window between 350 and 550 °C is suggested. Nevertheless the effect of different helium generation rates can be evaluated together with the determined influence of the irradiation temperature.

7. Discussion

7.1. Model evaluation

As shown in Chapter 6, the developed model is suitable to simulate helium bubble growth and calculate helium cluster size distributions for different helium production rates and for timescales relevant for high dose irradiation experiments as well as for fusion. The basic model was adapted to the irradiation experiments SPICE and ARBOR1 because experimental investigations on microstructural evolution have been made accessible by Transmission Electron Microscopy (TEM), providing a comprehensive study of helium bubbles and their effects on mechanical properties of the structural Reduced Activation Ferritic/Martensitic (RAFM) steel EUROFER97.

For a realistic description of helium bubble nucleation and growth, necessary simulation parameters were chosen close to the actual present material system: where possible, physical variables were taken for a Fe-10%Cr steel, else values for α-iron were used. Because calculation complexity especially for Molecular Dynamics (MD) simulations increases when considering an increasing number of different atom types, most simulations providing input data used in this work are restricted to α-iron. All relevant parameters could be found temperature dependent from literature studies, which was necessary for the description of the different irradiation experiments, and also for the conditions in fusion. Specifications of damage generation and gas production from BOR60 and High Flux Reactor (HFR) irradiation were exactly reproduced from literature descriptions by implementing necessary parameters into the model.

The model itself uses Rate Theory to describe helium bubble nucleation and growth. A set of rate equations on a discrete helium atom basis is solved for all bubble sizes by a self-developed Fortran code, which is shown in parts in Appendix B. Calculation times increase with increasing bubble sizes because the number of equations per solving step correlates with the number of the largest existing bubble size. The bubble size distribution shown in Fig. 6.12b for a temperature of 711 K yields a peak bubble diameter of 16.6 nm, which correlates to a bubble size of 198,117 atoms. Calculations though are performed up to bubble sizes of 277,000 atoms to assure a smooth density decrease for large bubbles to zero avoiding numerical instabilities. Even though the calculation time is large, it is possible to describe these irradiation conditions up to simulation times of one year. Nevertheless, for convenient calculation times the peak bubble diameter should stay below 10 nm (corresponding to 43,738 atoms) using the recent model configuration. Standard simulation, e.g. ADS3 under ARBOR1 irradiation (3.98×10^7 s) shown in Fig. 6.5, takes a calculation time of 17 hours yielding a peak bubble diameter of 7.9 nm corresponding to a bubble size of 21,488 atoms. Additional modifications, however, or corrections of numerical errors cause further increase in calculation time.

In Section 3.1.1 a scheme of the evolution of characteristic concentrations relevant for bubble nucleation and growth is presented (Fig. 3.2) which was derived from [61]. A comparison with simulation results of this work (Figs. 6.2 and 6.3) shows agreement in overall curves' behavior. Incubation, nucleation and growth regimes during irradiation can be clearly distinguished. The evolution of the monomer concentration and cluster density shows the same trend. Peaks of nucleation rate and monomer concentration coincide for all simulations. The condition for diatomic nucleation is satisfied because furthermore the helium amount bound to clusters intersects the monomer concentration at its peak position. The description of the annealing part in Fig. 3.2 is not reproduced by the model, reasons and further restrictions are explained in the following.

Limitations of the model are given by some of the assumptions made. Only helium monomers are considered to be mobile in the matrix. Therefore helium cluster migration and coalescence is not taken into account and possible coarsening mechanisms as described in [61] causing the cluster density decrease in Fig. 3.2 are reduced. A second coarsening mechanism, Ostwald ripening, is described. In this process a diffusion fluence is directed from smaller bubbles to larger ones, because differences in the chemical potential exist due to different curvature radii of the bubbles' surfaces, thus yielding higher gas pressures in smaller helium bubbles. Therefore a bubble size dependent helium resolution is observed, which can also be found in the description of the model's emission rate in Section 4.4.2. Coarsening would be relevant to describe especially at later stages of irradiation experiments when helium generation by boron transmutation ceases. Only little change in cluster size distributions can be observed between 10^7 s and the end of the SPICE irradiation program at 6.67×10^7 s, because the contribution of the helium generation by steel matrix atoms is small, yielding only 7 appm helium for this long irradiation period. By contrast, transmutation of helium by matrix elements gains importance the lower the boron transmutation rates and associated total helium amounts are, as in the case of ADS2 under ARBOR1 irradiation.

Vacancies are not explicitly described in the model. Generally, vacancies should be treated as a separate species, which itself undergoes diffusion and also clustering processes. However, it is assumed that due to low temperature irradiation the concentration of vacancies in the matrix is supersaturated from the very beginning of the irradiation. Therefore vacancies are always accessible for clustering of helium atoms, which exist as substitutional helium due to their low matrix solubility. The diffusion of vacancies is indirectly described by taking into account a vacancy mediated helium diffusion mechanism, where diffusivities were derived from kinetic Monte Carlo (kMC) simulations with values close to the helium-divacancy diffusion mechanism. Calculations of helium bubble growth, in this work, are performed for fixed helium-to-vacancy ratios in the bubbles, which can be adjusted at the start of the simulation. Due to the reasons mentioned in Section 4.1 values of the helium-to-vacancy ratio x_{He} are varied between 0.25 and 1. The assumption of a constant x_{He} for all bubble sizes is generally not valid, because smaller bubbles own a higher equilibrium pressure according to Eq. (4.27), which thus implies a higher helium density x_{He} inside the bubble. Large bubbles, on the other hand, may enter the void growth regime, where cavities grow by vacancy capturing alone, therefore leading to a decrease in x_{He} the larger the bubbles are. Nevertheless, since TEM investigations on EUROFER97 and boron doped variants did not show a bimodal bubble size distribution, the assumption is still a good approximation for the description of bubble growth at the stage preceding bubble-to-void transformation, allowing the assessment of the effects of x_{He} on simulated bubble size distributions.

In addition to the performed simulations in Chapter 6, an estimation of the influence of a size dependent helium-to-vacancy ratio on simulated bubble size distributions is presented in the following. Using a simple approach, $x_{He}(i)$ was derived by using a fitting curve as shown in Fig. 7.1a. Taking into account results from simulations and experiments presented in Chapter 3, a size dependent helium density was approximated by $x_{He}(i) = 1.21(1+i)^{-0.16}$. Thereby, emphasis was put on a helium-to-vacancy ratio close to 1 for small clusters (up to ten helium atoms), and $x_{He}(i)$ being 0.33 for a $\sim$4 nm sized bubble (as presented in Section 3.4.3.2). Even though the fitting curve yield a helium density of 1.08 for

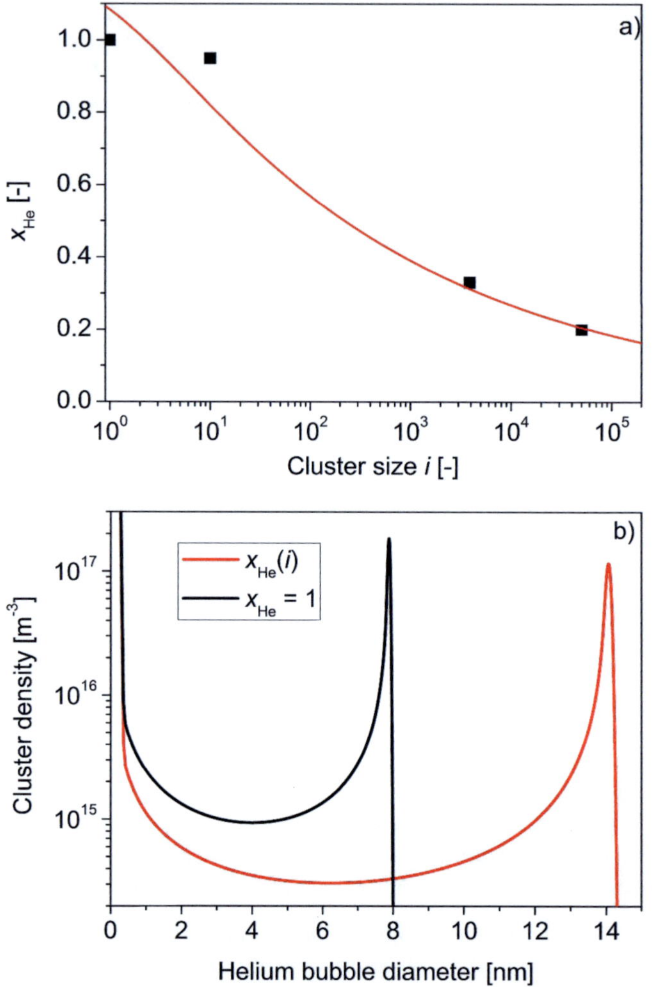

Figure 7.1.: Evaluation of a variable helium density x_{He} depending on cluster size i. a) Data points are estimated and partly based on results from simulations and experiments presented in Chapter 3. A fitting curve was used for simulations described by $x_{He}(i) = 1.21(1+i)^{-0.16}$. b) Comparison of standard simulations for ADS3 after ARBOR1 irradiation using a constant ($x_{He} = 1$) and size dependent ($x_{He}(i)$) helium-to-vacancy ratio.

a cluster size of one, it's curve evolution towards large cluster sizes is believed to be convenient for the simulation purpose. Eq. (4.7) for the cluster radius was adapted to

$$R_i = \sqrt[3]{\frac{3\Omega i}{4\pi x_{\mathrm{He}}(i)}} = b_i \sqrt[3]{i}, \tag{7.1}$$

and hence was used for the simulation. A comparison with the standard bubble size distribution for ADS3 after ARBOR1 is presented in Fig. 7.1b. The peak bubble diameter is shifted to 14.1 nm when using the size dependent helium density, but aside from a slightly broader peak, the general shape of the curve is comparable to the standard simulation. Further simulation attempts should use the appropriate equation of states for helium bubbles, e.g. as it was recently proposed in [122], resulting in a more accurate and physically based size dependent helium density.

Emission rates of helium atoms from existing clusters depend on helium solubility in the matrix and surface energy. Due to the low temperature irradiation conditions with helium trapped as a substitutional atom, substitutional helium solubility is used to calculate emission rates. Although substitutional solubility is much higher than the interstitial one (see Section 4.5), the resulting equilibrium concentrations are still very low (see Tab. 6.2). Nevertheless, the parametric study (as it was shown in Fig. 6.7) proved that a variation of the solubility C_1^{eq} has an influence on the simulated bubble size distribution, although its mean value was only 1.61×10^3 m^{-3}. The reason why emission rates play an important role in the presented model is due to the surface energy, whose exponential influence on the emission rate is responsible for the change in bubble size distributions. Since temperature dependent helium solubility is calculated by using ideal gas approximation, possible uncertainties are parametrically studied in a first attempt by increasing and decreasing solubility by one order of magnitude while the surface energy is kept constant. Peak bubble diameters were achieved for ADS3 in ARBOR1 of 10.1 and 6.6 nm, respectively, while standard values yield 7.9 nm. It has to be noted that an attempt was made to evaluate and calculate helium solubility on the basis of the chemical potential of a real gas according to equations given in [101], but the main calculation parameters (e.g. the chemical potential of an infinitely dilute interstitial solution) were not accessible. Therefore a comparison of both values with the corresponding difference can not be given.

Concerning the surface energy γ_{SF} values are determined as a function of temperature as derived from [100] for α-iron. Calculated values for SPICE and ARBOR1 relevant temperatures are 2.332 and 2.318 J/m^2, respectively. Most publications in literature assume values for γ_{SF}, which lie between 1 and 2 J/m^2. However, these values are assumed to be temperature independent, and their origin is not exactly presented. As it was shown in Section 6.3.2, the surface energy has a strong influence on the emission rate and thus the simulated bubble size distributions. Since the model should cover a wide range of irradiation conditions up to high temperatures expected in fusion, a temperature dependent description of the surface energy is assumed to be most accurate.

7.2. Comparison of simulations with microstructural investigations

7.2.1. Assessment of measured helium content by TEM

Bubble size distributions from Fig. 3.13 measured for ADS3 after ARBOR1 irradiation are re-evaluated. The purpose is to compare total helium amounts found by microstructural investigations with expected helium from simulations. For ADS3 after ARBOR1 irradiation, 50 appm helium is produced based on calculations shown in Fig. 6.1. Mean diameters from TEM investigations are re-calculated in terms of helium atoms i, using Eq. (4.7) with lattice constant values for room temperature. Tab. 7.1 shows resulting mean bubble sizes still depending on the helium-to-vacancy ratio x_{He}. The mean bubble size is built for all four areas: $\langle \bar{i} \rangle = 128.2\, x_{He}$. Taking the atomic volume calculated for room temperature from Eq. (4.4), the helium concentration in the investigated TEM sample can be estimated to 138.5 x_{He} appm. Assuming a constant helium-to-vacancy ratio, and

Table 7.1.: Re-calculation of mean bubble size from Fig. 3.13 in terms of helium atoms depending on helium-to-vacancy ratio x_{He}.

Area	Mean size $\langle i \rangle$ [helium atoms]	Total bubble density N [m^{-3}]
1	249.3 x_{He}	1.9×10^{22}
2	75.5 x_{He}	4.8×10^{22}
3	212.2 x_{He}	9.7×10^{21}
4	89.5 x_{He}	1.5×10^{22}

using the Electron Energy Loss Spectroscopy (EELS) result on the 4 nm sized bubble shown in Fig. 3.12 ($x_{He} = 0.33$), the helium amount in the sample equals 46 appm, which agrees well with simulation presettings of 50 appm. Furthermore this implies that most of the generated helium is observable and situated in visible bubbles, and about 8% in smaller bubbles under the detection limit of TEM. However, since the equilibrium bubble pressure increases for decreasing bubble sizes, the helium-to-vacancy ratio may also increase, and the detected helium amount will approach closer to the expected values. Nevertheless, due to continuous helium production, there still exists a margin in helium concentration reserved for small clusters and bubbles not visible in TEM, which are reported by simulations.

7.2.2. Analysis of helium bubble size distributions

At the start of the modeling activities of this work microstructural investigations were accessible for EUROFER97, ADS2 and ADS3 only after SPICE irradiation, and no quantification had been done on detected bubble size distributions. Qualitatively, bubbles had been observed up to sizes of 5 nm for ADS3 after SPICE irradiation [29, 30]. Therefore, the first simulation results for the SPICE experiment shown in Fig. 6.5 yielded reasonable bubble sizes.

Quantitative analyses on helium bubbles for EUROFER97 based boron doped model alloys are only available for ADS3 after ARBOR1 irradiation, as presented in Fig. 3.12. Simulation results are adapted by converting calculated continuous bubble size distributions into histograms with the same bin size of 0.25 nm like the data obtained by TEM. Fig. 7.2 shows resulting histograms for averaged TEM results compared to standard simulations for ADS3 after ARBOR1 irradiation ($x_{He} = 1$) and for a simulation with a helium-to-vacancy ratio of 0.5. Simulated histograms differ from previously presented size distributions e.g. from Fig. 6.5: histograms show a continuous increase in bubble density from sub-nanometer sized bubbles up to the peak diameter. The reason for this is the constant histogram's bin size, whereby intervals at small bubble sizes include less different bubbles when compared to regimes at larger bubble sizes due to the non-linear relation between diameter and containing helium atoms as given by Eq. (4.7).

Obviously, peak bubble diameters from TEM investigations and simulations for ADS3 after ARBOR1 irradiation do not coincide. The basic model using standard parameters overestimates peak bubble diameters achieved under ARBOR1

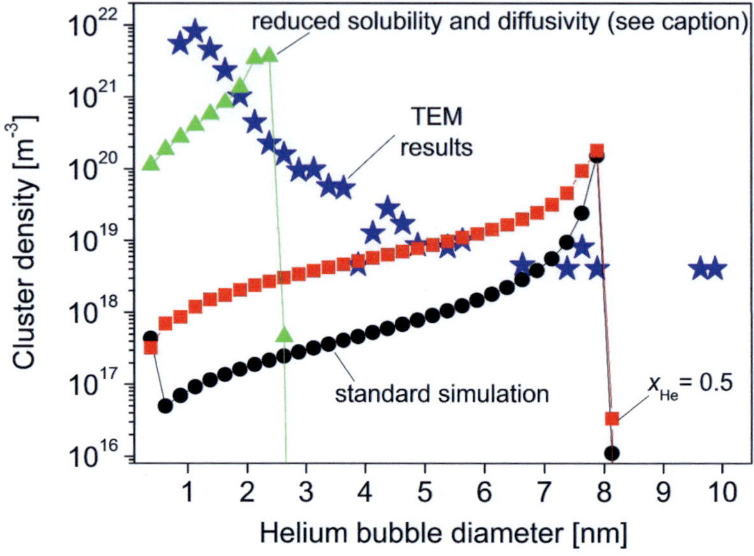

Figure 7.2.: Comparison of the bubble size distribution from TEM measurements with simulations [31]. TEM results ($\bigstar$) are averaged for all four areas shown in Fig. 3.13. Simulated size distributions are converted into histograms with a bin size of 0.25 nm according to TEM histograms. Size distributions show standard simulations ($x_{He} = 1$) for ADS3 after ARBOR1 irradiation ($\bullet$) and for a helium-to-vacancy ratio of 0.5 ($\blacksquare$). Additionally, helium solubility and diffusivity were decreased by one and three orders of magnitude, respectively, with $x_{He} = 1$ ($\blacktriangle$).

irradiation conditions. For this irradiation time of 3.98×10^7 s, the comparison of simulations for a helium-to-vacancy ratio x_{He} of 0.5 and 1 yield the same peak bubble diameters, but for $x_{He} = 0.5$ the density of smaller bubbles is increased and more comparable to TEM results. However, further decreasing of x_{He} to 0.25 will lead to larger peak bubble diameters under the same irradiation conditions, as it was shown in the parametric study in Section 6.3, and thus will not yield better agreement. Statistics for larger bubbles (> 4 nm) classified by TEM are quite poor, since a density of 4×10^{18} m^{-3} correlates only with one detected bubble within a counting range. That means, investigating larger volumes by TEM should decrease densities of larger bubbles when compared to bubbles at the peak diameter, but measurements will be very time consuming and the investigated volume of $0.196\,\mu m^3$ had already been quite large for the aim of detecting nanometer sized bubbles.

Taking the dependencies determined from the parametric study a reduction of both the effective helium diffusivity and solubility will shift the peak bubble diameter to smaller bubble sizes. Considering a reduction of both values by one order of magnitude, simulations for a helium-to-vacancy ratio of one yield for ADS3 after ARBOR1 irradiation a peak bubble diameter of 4.5 nm (see Fig. A.1). However, a peak diameter of 1.1 nm as obtained by TEM is hardly attainable when staying within reasonable parameter values. For comparison, Fig. 7.2 also shows a histogram which was recalculated from the second simulation shown in Fig. A.1 (green curve). Therein, helium solubility and diffusivity were reduced by one and three orders of magnitude from standard values, respectively. The resulting peak bubble diameter yield 2.3 nm, the corresponding cluster density is comparable to experimental TEM results. A further decrease of the helium solubility and diffusivity to match the experimental results is thought to be inappropriate because the difference from the underlying fundamentals would be too large. Furthermore, simulations for SPICE and ARBOR1 irradiation have to use the same physical basis, and changes of model parameters should affect simulations for both conditions in the appropriate way.

In terms of the curves' shape, simulations are not capable of reproducing the measured TEM bubble size distribution. Although the simulations yielding a smaller peak bubble diameter (as presented in Fig. 7.2) show a broadened peak, no simulation allows to describe the broad and slowly size dependent shoulder on the right side of the experimentally determined bubble size distribution. While TEM analysis on helium bubbles in ADS3 after ARBOR1 irradiation (see Section 3.4.3.2) did not detect a bimodal bubble size distribution, a bubble-to-void transformation of cavities is well accepted by means of theoretical considerations and was shown to occur under similar conditions in further microstructural investigations (see Section 3.5). The broad shoulder towards larger bubble sizes could thus be an indication of the onset of a bubble-to-void transformation, where several bubbles had reached the critical size. They might have started to grow as voids generating the isolated larger cavities detected by TEM.

A second reason for the curve mismatch of the simulated and experimental bubble size distribution could be related to the nucleation mechanism. Besides homogeneous di-atomic nucleation, which was described to be the main nucleation mechanism in this work, further nucleation paths assisting the decay of a

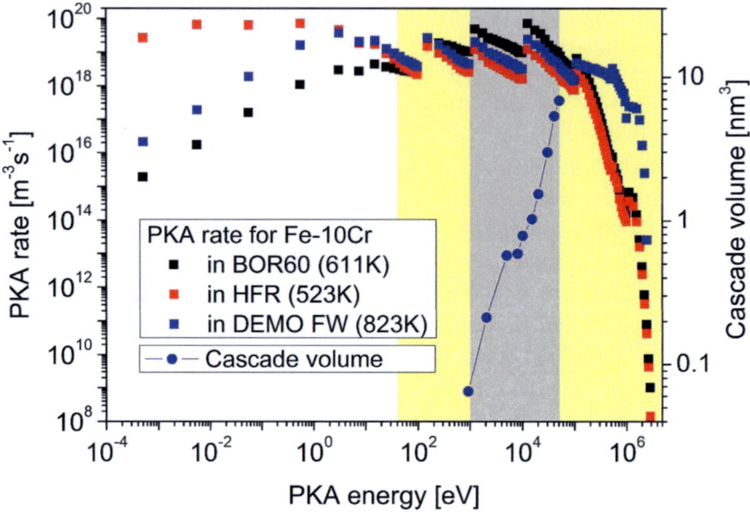

Figure 7.3.: Energy distribution of PKA rates in Fe-10%Cr calculated by [88] for specific neutron spectra from BOR60, HFR and FW of DEMO reactor. PKAs from the yellow energy range ($> 40\,\text{eV}$) are supposed to initiate displacement cascades. For PKA energies between $1\,\text{keV}$ and $50\,\text{keV}$ MD simulations [123] yield corresponding cascade volumes at the cascade's peak time.

supersaturated solid solution seem possible. Microstructural defects, whether they stem from interaction with the neutron irradiation or exist beforehand, may serve as nucleation centers for helium bubbles. Although TEM investigations on ADS3 after ARBOR1 irradiation shown in Section 3.4.3.2 reveal an overall homogeneous bubble distribution, it might be possible that the few larger bubbles detected could have been initiated by microstructural defects. This kind of heterogeneous nucleation, however, seems to have only minor influence on the overall experimentally determined bubble distribution, at least at this low irradiation temperature. Nevertheless, afterwards in Section 7.4 results of simulations of helium bubble nucleation and growth in the bulk and at grain boundaries (GBs) are superimposed, clearly demonstrating the possible effect on experimentally determined bubble size distributions.

Model evaluation yields a good agreement of the peak bubble diameters with experimentally determined qualitative results for SPICE, but simulations overestimate the achieved bubble sizes for ARBOR1. Therefore, the irradiation experiments SPICE and ARBOR1 have to be examined in detail to find a solution for the discovered inconsistency. For SPICE, helium generation by boron transmutation ceases already after 10^7 s, afterwards only transmutation from additional alloying elements enhances the helium concentration until the end of the experiment. Because the model is not capable of describing bubble coalescence, changes in bubble size distribution within this time period are very small. Higher temperatures in ARBOR1 (see Section 6.3) as well as a constant (even though lower) helium generation rate lead to large bubble sizes. The main difference between both irradiation experiments is given by different damage production rates: for ARBOR1, G_{dpa} is 5.62×10^{-7} dpa/s, which is 2.75 times the value of the SPICE specifications. In the basic model, G_{dpa} is only relevant for calculating helium generation rates, while further dependencies are not described. However, in Section 4.7 an extension of the model describing helium resolution from bubbles due to cascade-bubble interaction was presented. Although this modification and its influence on simulations was only investigated generally for a specific irradiation condition – namely ADS3 under ARBOR1 irradiation – still using a fitting parameter for the correlation of damage rate and affected bubble concentration, it clearly allows the following conclusions:

- Bubble sizes are significantly reduced by taking into account a constant helium resolution from all bubbles of 5% resulting in a peak diameter of 2.9 nm. Based on MD simulations shown in Section 3.6, a more realistic approach was described by a helium resolution depending on bubble size, whereby the simulation after one year of irradiation yields a peak diameter of 5.3 nm.

- Since the bubble density which is affected by displacement cascades directly correlates with the damage production rate G_{dpa}, the effect of helium resolution is expected to be more pronounced for the ARBOR1 experiment.

To find a physical basis for the correlation factor X_{corr} linking the affected bubble concentration with the damage production rate (see Eq. (4.32)), the following considerations are made. Neutrons impact on the sample volume causing the creation of primary knock-on atoms (PKAs) and the formation of displacement cascades. Fig. 7.3 shows the energy distribution of the PKA rate for the specific irradiation conditions of ARBOR, SPICE and First Wall (FW) of future fusion DEMO reactor. Data was calculated by [88] using SPECTER code [124] for neutron spectra from BOR60, HFR and DEMO FW [125]. From PKA rates the

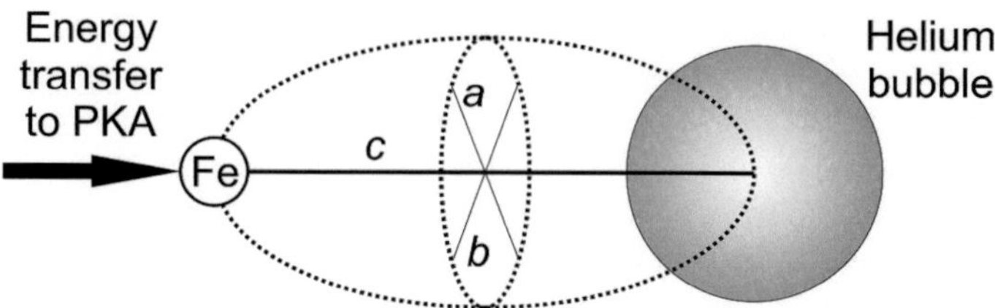

Figure 7.4.: Scheme of a displacement cascade with an ellipsoidal cascade volume initiated at a PKA with distance c to the bubble center.

affected volume per PKA can be estimated by taking into account calculations from [123], where cascade peak volumes were derived from MD simulations, but only for PKA energies between 1 keV and 50 keV. Further results from [88] can be used, because in addition to the results shown in Fig. 3.14 distances of PKAs to the bubble centers of up to 20 times the lattice constant a_{Fe} were used. It was shown that with increasing PKA distance the helium resolution from the simulated bubble decreases, and for distances larger than 8.5 times (15 times) the lattice constant no helium resolution could be observed (see Fig. A.2) for 5 keV (10 keV) cascades. Assuming an ellipsoidal cascade volume with mentioned PKA distances as c axis and a and b axes with half the expansion as shown in Fig. 7.4, the cascade volumes can be estimated to 0.6 and 3.2 nm^3 for 5 and 10 keV cascades, respectively. While the former value fits exactly the calculation from [123], the latter is too large. However, since MD simulations were performed only for a limited number of calculations, statistics should be improved. Additionally, a mean effective distance (volume) has to be defined, within which bubbles are affected by the PKA, and also the shape of the displacement cascade needs to be clarified. Nevertheless, this method will allow assessment of the volumes affected by cascades of different PKA energies per time period and therefore the possible effects on growing helium bubbles.

7.3. Assessment of helium induced hardening

Assessment of helium induced hardening was performed for simulated bubble size distributions using the Dispersed Barrier Hardening (DBH) model. Simulated hardening results from Section 6.7 shown in Tab. 6.5 are compared to experimentally determined hardening from [79, 80]. Fig. 7.5 presents corresponding hardening values specific for ADS2 and ADS3 after SPICE and ARBOR1 irradiation. The variation in experimental values stems from the measuring inaccuracy, while the shaded bars are due to the variation of the barrier strength α of helium bubbles, which is assumed to lie between 0.2 and 0.4.

The calculated hardening is in good agreement with an experimentally observed extra hardening for both boron doped steels in both irradiation experiments. While simulated helium bubble size distributions alone would cause the hardening calculated in Tab. 6.5, superposition effects with other obstacle types have to be considered. The direct comparison of helium hardening estimated on the basis of simulated bubble size distributions with experimentally determined extra hardening will therefore imply a linear superposition law of hardening contributions between helium bubbles and other obstacle types. As mentioned in Section 3.4, hardening due to helium only shows a minor contribution to the overall measured hardening. The largest impact on hardening is caused by dislocation loops, while further defects e.g. chromium rich α' precipitates also might play a role. In the case of the validity of a root sum square (RSS) superposition law [127] the calculated helium hardening strongly underestimates the experimentally observed extra hardening by taking into account irradiation induced (non-helium) hardening of base EUROFER97 steel of 405 MPa. A comparison of both calculation methods for standard simulations of ADS3 after ARBOR1 irradiation is given in Tab. 7.2. Helium induced hardening ($\alpha = 0.4$) decreases from 27.2 MPa, when treating the helium bubble size distribution as the only occuring hardening cause, to effective 0.9 MPa, when superimposing simulated helium hardening with experimental non-helium hardening of 405 MPa considering the RSS superposition law.

Furthermore, Tab. 7.2 shows calculated hardening with a barrier strength of $\alpha = 0.4$ after one year of irradiation taking into account other different simulation configurations of ADS3 in ARBOR1. Simulations for $x_{He} = 0.25$ yield the highest helium induced hardening of all calculations, standard simulations the lowest. Considering an attachment barrier of $E_i = 0.17$ eV in the capture rate also leads to an increase in calculated helium hardening. The model modification of helium resolution from bubbles is most promising because not only does it reduce peak diameters of the bubble size distributions, it also increases helium induced hardening. Both effects achieve results approaching closer to experimental investigations. As mentioned before, the usage of an RSS superposition rule for

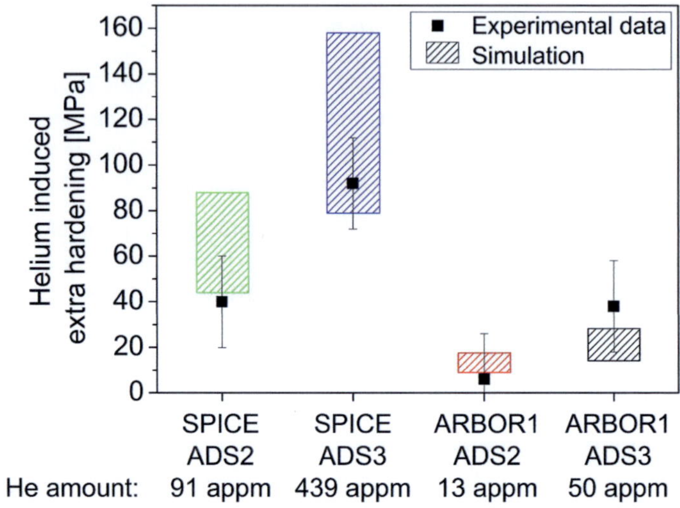

Figure 7.5.: Comparison of experimentally determined helium induced hardening from [79, 80] with theoretical results applying the DBH model on simulated bubble size distributions using standard parameters [126]. The range of the simulated hardening values stems from varying the barrier strength α from 0.2 to 0.4.

calculating effective helium induced hardening results in hardening values too low for experimental observations. However, the effective superposition should be chosen as a mixture of linear and RSS superposition depending on the actual defect types, densities, etc. as it is proposed in [127]. Therefore the calculated values should be taken as upper and lower limits of helium induced hardening.

7.4. Sink effects

An interesting point is given by the description of the helium losses to sinks. While the helium monomer concentration in the matrix is lowered by the fraction of helium atoms diffusing to sinks, the helium is not lost, but stored at the different sink sites. Therefore, depending on the sink strengths, the sinks themselves may act as nucleation sites for further helium clustering. While the creation of helium bubbles described by the model is based on homogeneous nucleation in the

Table 7.2.: Application of the DBH model ($\alpha = 0.4$) on different simulated bubble size distributions after one year of irradiation. Based on standard simulations for ADS3 in ARBOR1, the mentioned parameter in the left column was altered for each case. Helium induced hardening for each simulation is calculated. By using a RSS superposition law, effective helium hardening is calculated taking into account the (non-helium) hardening of base EUROFER97 steel (405 MPa).

Simulation Deviation from standard	He hardening [MPa] due to simulation	Eff. He hardening [MPa] Superposition of non-He hardening using RSS law
ADS3 in ARBOR1	27.2	0.9
$x_{He} = 0.5$	40.6	2.0
$x_{He} = 0.25$	44.2	2.4
$E_i = 0.17\,\mathrm{eV}$	39.9	2.0
Size dependent He resolution	38.4	1.8

matrix, clustering at sinks must be attributed to a heterogeneous type. Recalling the TEM micrographs of ADS3 after irradiation at a temperature of 450 °C to 18.1 dpa in SPICE [29, Fig.7] cited in Section 2.3, heterogeneous distribution of helium bubbles concentrated at sinks, e.g. line dislocations or grain boundaries, might be explained as follows: If the sink strength is strong enough, the main fraction of helium atoms will diffuse to the sinks and the remaining helium concentration in the matrix will be small, leading to a primary bubble nucleation at sink sites. Temperature plays a dominant role in this process, because of the strong temperature dependence of the helium diffusivity which influences the loss rate to sinks according to Eq. (4.30). That is why at higher temperatures a greater influence of heterogeneous nucleation is expected [61].

The two-dimensional (2-D) variant of the developed model presented in Section 4.8 completes the full description of the generated helium during irradiation, it steps in where the three-dimensional (3-D) model describing homogeneous clustering in the bulk ends. The time dependent helium concentration caught by GB sinks leads to bubble nucleation and growth at this microstructural defect. Under the presented assumptions the helium bubble size distribution in ADS3 after ARBOR1 irradiation is calculated in Section 6.6, where bubbles are homogeneously distributed over the whole area of GBs. The simulation result shown in

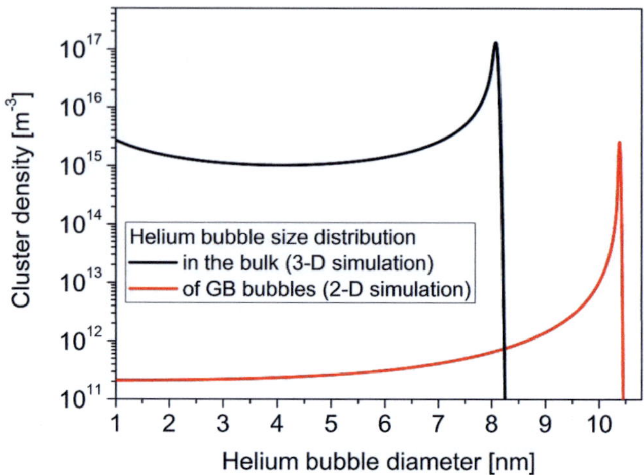

Figure 7.6.: Final helium bubble size distributions for ADS3 after ARBOR1 irradiation considering grain boundaries as sinks for helium. The 3-D simulation allowed helium losses to line dislocation and grain boundary sinks, whereof LDs caught only an insignificant amount of helium. For a direct comparison 2-D GB simulations from Fig. 6.17b were recalculated for a 3-D state.

Fig. 6.17b is now converted back by using Eq. (4.34), providing bubble densities in units of m^{-3}. Both simulated size distributions are presented in Fig. 7.6.

For the given irradiation conditions and sink parameters, 2.5% of the helium amount produced in ADS3 during ARBOR1 irradiation is lost to GB sinks (corresponding to 1.2 appm helium). Bubble densities formed at GBs are therefore by many orders of magnitude smaller than the ones calculated for the bulk. Nevertheless, peak diameters achieved at GBs are by 2.3 nm larger, and the peak density is only two orders of magnitude lower than the value in the bulk. Speculating on a microstructural quantitative investigation of bubbles within this specimen, at least the peak sized bubbles at GB sinks should be observable and play a role in the measured bubble size distribution. Since their distribution is not homogeneous over the whole sample volume, TEM measurements may allow a separate analysis which may provide different bubble sizes and densities for bulk and sink bubbles, leading to the adaptation of necessary simulation parameters for helium clustering at sinks. Furthermore, if a distinction between both size distributions is not made or cannot be made, an analysis will provide a bimodal bubble size distribution, with an observation of a small density of larger bubbles. Therefore not only a bubble-to-void transformation of existing cavities can be responsible for a bimodal bubble size distribution, but also microstructural defects like GBs.

Model modifications implemented already for the 3-D variant can also be applied to helium bubble nucleation and growth at GBs. Especially helium resolution from bubbles due to displacement cascade interaction seems to be a promising effect, which should be used after determining the appropriate simulation parameters. Emission of helium from GB sinks is not considered, and further investigations should evaluate the relevance of this mechanism. This would however require a backwards coupled solving of the 2-D and 3-D model, resulting in a more complex numerical code with high calculation times. The configuration of the simulated GB is assumed in an elementary way which allows fast calculation up to long simulated irradiation times covering irradiation experiments and fusion.

7.5. Comparison of irradiation experiments with fusion relevant conditions

Irradiation conditions in the programs SPICE and ARBOR1 differ from the expected fusion conditions. To gain high energy efficiency, the temperature in the FW of a fusion reactor has to be as high as possible to offer a large temperature range for the cooling medium, but it is limited by the maximum operation temperature of the steel. For EUROFER97, the upper application limit is given by the creep strength, which is acceptable up to $550\,^{\circ}$C [15]. Minimum temperatures should not fall below $350\,^{\circ}$C, because otherwise a large embrittlement is expected. In fusion power plants, dose rates between 20 and 30 dpa/year with helium generation rates from 10 to 15 appm/dpa are expected in the FW. Transmutation rates are assumed to remain unchanged with time due to a high fraction of helium producing isotopes, e.g. the isotope ^{54}Fe is enclosed at a percentage of 5.8% within iron and produces helium by an (n-α)-reaction. These conditions will result in helium concentrations of 200–450 appm/year. While in ARBOR1 a constant helium generation rate was achieved, the produced helium amount per year (40 appm for ADS3) is too low when compared to fusion conditions. The helium concentration after the first year of irradiation in SPICE (435 appm for ADS3) fits within the range of the expected helium amount under fusion neutrons, but the transmutation rate is not constant. While temperatures in ARBOR1 cover the lower regime of expected FW temperatures, the whole SPICE experiment took place within a temperature range of 250 and 450 °C. In comparison to ARBOR1 the diffusivity of helium is estimated to be two orders of magnitude higher at 550 °C (see Fig. 3.2.1).

Simulation results from Section 6.8 are shown in Fig. 7.7 and compared to other simulations with constant helium generation rates. Bubble size distributions after one year of irradiation are analyzed and peak bubble diameters and total cluster densities depending on the helium generation rates are shown in Fig. 7.7a and b, respectively. The influence of the effective helium diffusivity can be observed by comparing the black and blue curve. As all other parameters are kept constant, the decrease of the diffusivity to 12% of the initial value causes a decrease of the peak bubble diameter by 1.8 nm and an density increase by a factor of 3. Standard simulations for ADS2 and ADS3 under ARBOR1 irradiation are also included in the diagram (red circles) since their helium generation rates within the considered time regime of one year are almost constant. This data can directly be compared to the black curve, because the helium diffusivity is the same for both conditions as can be seen from Tab. 6.2, and only the other parameters (as shown in Fig. 6.6 of Section 6.3) are influenced by the higher temperature of the ARBOR1 experiment affecting simulation results. ARBOR1 irradiation is responsible for peak diameters which are by 1 nm larger than black curve values, which causes as a consequence a reduced total cluster density. Simulations for helium generation rates relevant to fusion are only performed with the parameter set of the black curve. As shown in Fig. 7.7 the peak diameter slightly increases (less than 1 nm) between a helium generation rate of 200 and 500 appm/year, while the total cluster density increases by a factor of 1.8. Both curves show a steep increase at low helium generation rates. For the peak diameter, the bubble growth seems to saturate with increasing helium generation rate. This effect is enforced due to the dependence of R_i from $\sqrt[3]{i}$, which leads to minor differences of radii between size adjacent bubbles at larger bubble sizes. Nevertheless, one observes also less increase of the peak bubble size based on i with higher helium generation rates. For the total cluster density, its increase with the helium generation rate is almost constant in the considered fusion relevant regime. The limited number of simulations with constant helium generation rate does not allow to extrapolate the red and blue curves towards larger values up to 500 appm/year. However, based

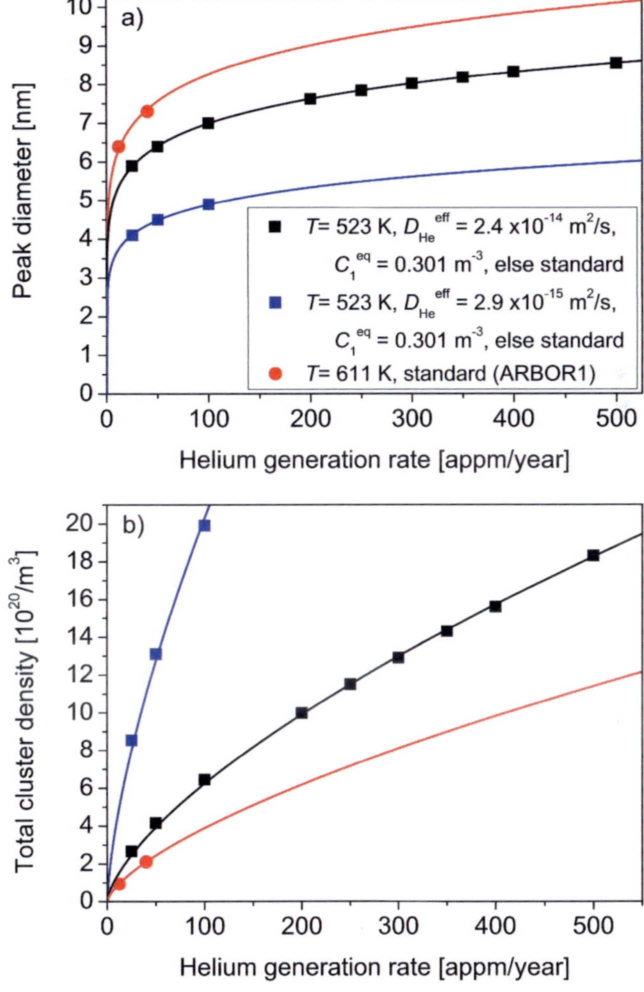

Figure 7.7.: a) Peak bubble diameters and b) total cluster densities for different helium generation rates after one year of irradiation. Generation rates up to fusion relevant values are considered for a specific parameter set. The influence of the helium diffusivity is presented. Data from simulations of ADS2 and ADS3 under ARBOR1 conditions is compared. Data was fitted by using a power law in the form of $y = Bx^c$ with fitting parameters shown in Tab. 7.3.

Table 7.3.: Parameters for fitting curves in Fig. 7.7 assuming a power law in the form of $y = Bx^c$. Subscripts *dia* and *dens* distinguish between fits for the peak bubble diameter and the total bubble density, respectively.

Parameter set	B_{dia} [nm]	c_{dia} [-]	B_{dens} [m^{-3}]	c_{dens} [-]
■	3.93	1/8	2.90×10^{19}	2/3
■	2.75	1/8	9.42×10^{19}	2/3
●	4.65	1/8	1.81×10^{19}	2/3

on the data points of the black curves in Fig. 7.7, a power law in the form of $y = Bx^c$ was used to fit both the peak bubble diameter and the total cluster density. For the peak bubble diameter, c_{dia} was set to $\frac{1}{8}$ for all three parameter sets, while for the total cluster density $c_{\mathrm{dens}} = \frac{2}{3}$. The curve dependent coefficients B_{dia} and B_{dens} are given in Tab. 7.3. Obviously, the data from simulations with varying parameter sets is well described by the fitting curves under the assumption of a constant power exponent. Nevertheless, further simulations are necessary with parameters following the application requirements for RAFM steels under fusion FW conditions to verify the proposed fitting curves.

8. Summary and outlook

The first part of this study was dedicated to the development of a physically based model to describe helium bubble nucleation and growth in Reduced Activation Ferritic/Martensitic (RAFM) steels under neutron irradiation. The model relies on Rate Theory using kinetic rate equations to describe the diffusion governed growth process. An adaptation of the model was performed to meet the specific irradiation conditions of different ^{10}B doped specimens in the irradiation experiments SPICE and ARBOR1. The necessity of using these experiments is given by the fact that up to now a fusion–like neutron spectrum cannot be reproduced in experiments. Helium production rates and contents have to be adjusted, e.g. by boron doping, to approach closer to fusion expectations. Therefore, model validation has to be done by accessing available experimental results. It was shown that in particular the ARBOR1 experiment is convenient for this purpose due to its constant helium generation rate over the whole irradiation time. Hence quantitative Transmission Electron Microscopy (TEM) investigations on boron doped RAFM steels were promoted especially on ARBOR1 specimens. Measurements on ADS3 have been made available during this study, and were used to assess crucial model parameters. All main parameters could be derived as a function of temperature from literature studies, provided by experimental investigations as well as by different simulation techniques.

In a second step, a program code was successfully compiled to numerically solve a large set of rate equations. Emphasis was put on a fast and efficient calculation to achieve simulated irradiation times in the range of the irradiation experiments and fusion. Furthermore, numerical stability of the calculations was ensured by the development and implementation of several analyzing and controlling program parts and subroutines, e.g. an automatic step size control. Some simulations took advantage of a self-written corrector code, which avoided numerical instabilities and kept the numerical error within an allowed accuracy. The code was steadily improved during model modifications because otherwise the more complicated model with more comprehensive calculations would have increased calculation times beyond acceptability.

Simulations yield the time dependent evolution of specific concentrations relevant for helium bubble nucleation and growth. It was shown that the model follows the conditions for diatomic nucleation. Incubation, nucleation and growth regimes can be clearly distinguished and appear as classic bubble growth theory predicts. The onset and evolution of the nucleation phase was shown to strongly affect the final bubble size distributions.

Final bubble size distributions for both boron doped alloys ADS2 and ADS3 after SPICE and ARBOR1 irradiation were determined with standard simulation parameters. The influence of the main model parameters on bubble growth were evaluated by performing a parametric study, furthermore the model and the numerical solving code was validated to provide reasonable results for all parameter sets used. The model was shown to predict peak bubble diameters for the SPICE experiment, which match preliminary existing qualitative microstructural investigations. However, the model overestimates peak bubble diameters for the ARBOR1 experiment, where quantitative measurements on helium bubbles in irradiated ADS3 specimens had been performed. The effect of helium resolution from bubbles due to cascade-bubble interaction was found to provide a partial solution to the problem of inconsistency between SPICE and ARBOR1 results. The higher damage rate in the ARBOR1 experiment enhances helium resolution from bubbles leading to a decelerated bubble growth and therefore smaller peak diameters. Further studies, e.g. by Molecular Dynamics (MD), should be performed on cascade-bubble interaction to allow the determination of the appropriate simulation parameters for this model extension.

The influence of sinks on helium bubble evolution was investigated. For the given parameters grain boundaries (GBs) were shown to own a higher sink strength than dislocations (DLs). Higher temperatures and lower helium generation rates enhance the fraction of the helium amount lost to sinks: the effect was most pronounced for ADS2 under ARBOR1 irradiation conditions. The description of the helium loss to sinks was completed by considering helium bubble nucleation and growth at GB sink sites. For this reason a two-dimensional (2-D) variant of the model was created which uses the captured helium concentration by sinks from the three-dimensional (3-D) model as input data. The adaptation of the model parameters to GB specifications yielded bubble size distributions evolving in the GB plane. A comparison of bubble size distributions was presented for bulk and

GB bubbles calculated for conditions of ADS3 under ARBOR1 irradiation. GB bubbles, which possessed only 2.5% of the total helium amount, achieved higher peak diameters. Although their density is lower, these bubbles are supposed to be observable by TEM. Further simulations should be performed at higher temperatures, to reveal changes in the relation between bulk and GB size distributions leading to primary bubble nucleation and growth at GBs corresponding to microstructural TEM investigations.

Helium induced hardening was determined by applying the Dispersed Barrier Hardening (DBH) model to the calculated cluster size distributions. For a size-invariant barrier strength simulated hardening was compared to experimental results. Good agreement was achieved by assuming a linear superposition of helium induced hardening from standard simulation (ADS3 after ARBOR1) with hardening from other obstacle types, while the usage of a root sum square (RSS) superposition law underestimates the helium effect. It was shown that model modifications, which lead to a reduced peak bubble diameter closer to experimentally measured bubble size distributions, also increase simulated helium induced hardening.

The model was proven to be suitable for simulating helium bubble growth under fusion relevant helium generation rates. For a given temperature, the evolution of the peak bubble diameter and the total cluster density was investigated when changing the helium generation rate. Results after one year of irradiation showed that the increase of the peak bubble diameter slows down with increasing helium generation rate, and the total bubble density almost linearly increases in the considered regime. Future investigations should provide a more comprehensive study at fusion relevant temperatures, determining the effects on helium bubble growth over the whole range of expected environmental conditions.

Concerning future modeling, the possibility of describing vacancies together with helium atoms in a combined and coupled clusterization approach should be pursued. A consequent continuation and enduring implementation of the mechanisms found to be crucial for helium nucleation and bubble growth, e.g. helium resolution from existing bubbles, should be performed, using e.g. additional simulation methods to determine missing and hitherto estimated modeling parameters. The usage of an appropriate real gas equation of state for calculating helium solubility in the matrix and helium density in the bubbles should improve the accuracy of the simulations provided that the necessary parameters can be derived. Besides helium induced hardening, which can be deduced in a simple way by applying a suitable hardening model to the determined bubble size distributions, the evaluation of helium induced embrittlement is more sophisticated, but should also be considered in future simulation approaches.

A. Experimental data and simulations

A.1. Mechanical properties

Table A.1.: Experimental results of hardening and embrittlement after the SPICE [79, 23, 22] experiment (σ_{ys} – yield strength from tensile tests; σ_{Dy} – calculated dynamic yield stress from impact tests).

	SPICE (250 °C, 13.6 dpa)		
	EUROFER97	ADS2	ADS3
DBTT (unirr.) [°C]	-90.8	-73.5	-99.8
DBTT (irr.) [°C]	58.2	197.5	270.2
σ_{ys} (unirr.) [MPa]	452.8±30.9 [a]	433.4±1 [a]	418.1±11 [a]
σ_{ys} (irr.) [MPa]	845.0±55.6 [a]	813.3±1.5 [a]	789.8±10 [a]
σ_{Dy} (unirr.) [MPa]	391	384	407
σ_{Dy} (irr.) [MPa]	796	829	904

$$^{a} - T = 300\,°C$$

Table A.2.: Experimental results of hardening and embrittlement after a) ARBOR1 [80, 81] and b) ARBOR2 [82] experiments (σ_{ys} – yield strength from tensile tests).

a)	ARBOR1 (330 °C, 32 dpa)		
	EUROFER97	ADS2	ADS3
DBTT (unirr.) [°C]	-90	-74 [b]	-100 [b]
DBTT (irr.) [°C]	107	174 [b]	174 [b]
σ_{ys} (unirr.) [MPa]	445.3	411.9 [c]	409.8 [c]
σ_{ys} (irr.) [MPa]	916.0	888.6 [c]	918.1 [c]

[b] – 22.4 dpa, [c] – 30.2 dpa

b)	ARBOR2 (330 °C, 69.8 dpa)		
	EUROFER97	ADS2	ADS3
DBTT (unirr.) [°C]	-90.8	-74	-100
DBTT (irr.) [°C]	136	238	260
σ_{ys} (unirr.) [MPa]	436.4±8.6	413.8±4.9	406.6±1.2
σ_{ys} (irr.) [MPa]	1106±110	1086.8±113.8	1134.8±134.8

A.2. Simulation results

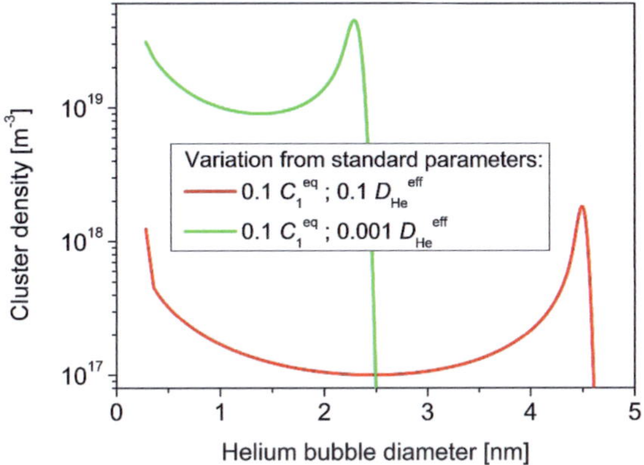

Figure A.1.: Simulation for ADS3 under ARBOR1 condition after an irradiation time of 3.98×10^7 s. Helium solubility was reduced by one order of magnitude to $C_1^{eq} = 1.61 \times 10^2$ m^{-3}, diffusivity by one ($D_{He}^{eff} = 2.4 \times 10^{-15}$ m^2/s) or three orders of magnitude.

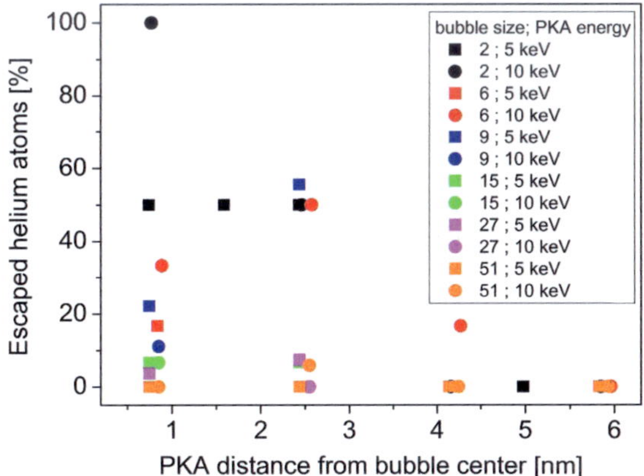

Figure A.2.: Helium resolution from bubbles of different sizes in α-iron for different distances of PKA to the bubble center. Simulations were performed by [88] using MD as shown in Section 3.6.

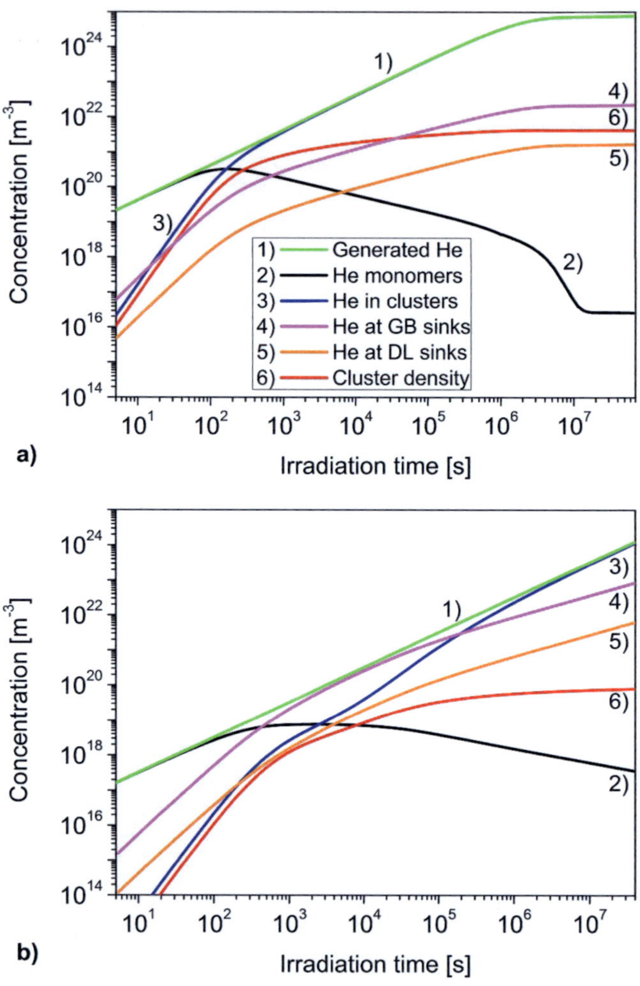

Figure A.3.: Comparison of concentration evolution for ADS2 in a) SPICE and b) ARBOR1 for $x_{\mathrm{He}} = 1$, taking into account grain boundaries (GB) and dislocations (DL) as helium sinks (see Section 6.4).

B. Fortran program code

B.1. Automatic adaptation of right boundary

```fortran
      ...
          if ((icalc.LT.imax).AND.(C(icalc-10).NE.0)) then
            icalc=icalc+100
            if (icalc.GE.imax)  icalc=imax
            write(*,*)  icalc
          endif
      ...
```

B.2. Automatic step size control

```fortran
! ================================================================
        subroutine stepsizecontrol(SSC,C,Cl,icalc,dt,k,g,aACC,rACC,
          tact)
        !automatic step size control
! ================================================================
        use omp_lib
        implicit none
        integer i,icalc,VAR,SSC,PROCS
        double precision dt,Nc0,Nc1,err,err2,aACC,rACC,tact,
     +  k(1:icalc),g(1:icalc),
     +  Cc0(1:icalc),Cc1(1:icalc),Cc2(1:icalc),C(1:icalc),
     +  Cl(1:icalc)
        data VAR/0/
        save VAR
!
        PROCS=OMP_GET_NUM_PROCS()-1
        if (PROCS.EQ.0)  PROCS=1
!
```

```fortran
18          Cc0(1)=C1(1)
19 !two times with dt/2
20 !$OMP PARALLEL DO num_threads(PROCS),shared(C,Cc1,k,g,icalc,dt)
21 !$OMP+    ,private(i)
22        do i=2,icalc-1
23          Cc1(i)=C(i)+0.5*dt*(-k(i)*C(1)*C(i)-g(i)*C(i)+
24     +               k(i-1)*C(1)*C(i-1)+g(i+1)*C(i+1))
25        enddo
26        Nc0=C1(1)
27 !$OMP PARALLEL DO num_threads(PROCS),shared(Cc0,Cc1,Nc0,k,g,icalc,
        dt)
28 !$OMP+    ,private(i)
29        do i=2,icalc-1
30          Cc0(i)=Cc1(i)+0.5*dt*(-k(i)*Nc0*Cc1(i)-g(i)*Cc1(i)+
31     +               k(i-1)*Nc0*Cc1(i-1)+g(i+1)*Cc1(i+1))
32        enddo
33 !
34 !second step with dt
35        Nc1=C1(1)
36 !$OMP PARALLEL DO num_threads(PROCS),shared(Cc1,C1,Nc1,k,g,icalc,
        dt)
37 !$OMP+    ,private(i)
38        do i=2,icalc-1
39          Cc1(i)=C1(i)+dt*(-k(i)*Nc1*C1(i)-g(i)*C1(i)+
40     +               k(i-1)*Nc1*C1(i-1)+g(i+1)*C1(i+1))
41        enddo
42 !one step with 2*dt
43 !$OMP PARALLEL DO num_threads(PROCS),shared(Cc2,C,k,g,icalc,dt)
44 !$OMP+    ,private(i)
45        do i=2,icalc-1
46          Cc2(i)=C(i)+2*dt*(-k(i)*C(1)*C(i)-g(i)*C(i)+
47     +               k(i-1)*C(1)*C(i-1)+g(i+1)*C(i+1))
48        enddo
49 !
50 !assessment and adaption of dt
51        if(sum(C).EQ.0) goto 300
52        !
53        do i=1,icalc
54        err=aACC+max(C(i),C1(i))*rACC
55        if(ABS(C1(i)-Cc0(i)).GT.err)then
56        dt=0.5*dt
57        VAR=1
58        SSC=1
59        goto 300
60        endif
61        enddo
62        !
```

```fortran
63          if (VAR.EQ.0) then
64 !$OMP  PARALLEL DO  num_threads (PROCS)
65 !$OMP+    ,shared (icalc ,aACC,C,C1,rACC,Cc1,Cc2)
66 !$OMP+    ,private (err,i),  reduction (+:VAR)
67        do i =1,icalc
68        err=aACC+max (C( i) ,C1( i))*rACC
69        if (ABS(Cc1 ( i )−Cc2 ( i )) .GT. err ) VAR=VAR+1
70        enddo
71        !
72        if (VAR.EQ.0)  then
73        dt=2* dt
74        !
75        if ((dt −1.GT.0) .AND. (tact −10000.LT.0)) then          !maximum dt
                   of 100  seconds
76        dt=1
77        goto  400
78        elseif (dt −2E2.GT.0)  then
79        dt=2E2
80        goto  400
81        endif
82        !
83        SSC=1
84        goto  300
85        endif
86        endif
87  400 continue
88        VAR=0
89        SSC=0
90 !
91 ! ================================================================
92  300 return
93        end
94 ! ================================================================
```

B.3. Correction of monomer concentration

```fortran
1 ...
2        IF (C1_FIT_CORR .EQ. 1 )THEN
3        c1_content =0.0
4        do i =2,icalc                          !sum of i*C( i , t* dt) for i
                   >=2 ——> He atoms bound in clusters
5           c1_content=c1_content+i*C1( i )
```

```fortran
 6              enddo
 7              !
 8          if (MOD( t ,2 ) .NE.0)  goto  302
 9            if (( tact −t_corr .GT.0) .AND.
10    +                    (ABS(C_He−cl_content )/(1E−3*C_He) .GT. 1 ) ) then
11          if (C_He−cl_content .GT.0)  C(1)=C(1) *(1+1E−2)
12          if (C_He−cl_content .LT.0)  C(1)=C(1) *(1−1E−2)
13          GOTO 300
14          endif
15          ENDIF
16  ...
```

B.4. Cascade induced helium resolution from bubbles

```fortran
 1  ...
 2          IF (BUB_RESOLUTION .EQ. 1 )THEN
 3          !
 4          do  i =1 ,icalc
 5             resol_size (i )=i *0.95
 6             if ( resol_size ( i ) .EQ.0)  resol_size (i )=1
 7          enddo
 8          continue
 9          !
10  !$OMP PARALLEL DO  num_threads (PROCS) , shared (C, icalc ) , private ( i )
11          do  i =2 ,icalc
12              C( i )=0
13          enddo
14          !
15  !$OMP PARALLEL DO  num_threads (PROCS)
16  !$OMP+          , shared (C, icalc , resol_size , dt ,Gdpa,C1) , private ( i )
17          do  i =2 ,icalc
18              C( resol_size (i ))=C( resol_size ( i ))+dt *Gdpa*C1( i ) *50
19              C( i )=(1− dt *Gdpa*50) *C1( i )
20          enddo
21          !
22  !$OMP PARALLEL DO  num_threads (PROCS)
23  !$OMP+          , shared (C, icalc , resol_size , dt ,Gdpa,C1) , private ( i )
24          do  i =2 ,icalc
25              C(1)=C(1)+C1( i ) *dt *Gdpa*(i−resol_size ( i )) *50
26          enddo
```

```fortran
27  !
28          ELSE
29          !
30  !$OMP  PARALLEL DO num_threads(PROCS),shared(C,C1,icalc),private(i)
31          do i=2,icalc-1
32            C(i)=C1(i)
33          enddo
34          C(icalc)=0
35          C1(icalc)=0
36          !
37          ENDIF
38  ...
```

References

[1] 2010 Survey of Energy Resources. World Energy Council, 2010. ISBN: 0946121265.

[2] World Energy Outlook 2010. International Energy Agency, 2010. ISBN: 9789264086241.

[3] K. Ikeda. ITER on the road to fusion energy. *Nuclear Fusion*, 50(1):014002, 2010.

[4] ITER Organization. ITER - The Way to New Energy. online. http://www.iter.org/.

[5] M. Rieth, M. Schirra, A. Falkenstein, P. Graf, S. Heger, H. Kemp, R. Lindau, and H. Zimmermann. Eurofer97 - tensile, charpy, creep and structural tests. *Wissenschaftliche Berichte FZKA*, 6911:1–83, 2003.

[6] NASA. Sun Fact Sheet. online. http://www.nasa.gov.

[7] C. E. Rolfs and W. S. Rodney. *Cauldrons in the Cosmos - Nuclear Astrophysics*. The University of Chicago Press, 1988.

[8] G. W. C. Kaye and T. H. Laby. *Tables of Physical and Chemical Constants*. Longman, www.kayelaby.npl.co.uk, 16th edition, 1995. ISBN: 0582226295.

[9] A. M. Bradshaw, T. Hamacher, and U. Fischer. Is nuclear fusion a sustainable energy form? *Fusion Engineering and Design*, 86:2770–2773, 2011.

[10] E. Roos and K. Maile. *Werkstoffkunde für Ingenieure - Grundlagen, Anwendung, Prüfung*. Springer-Lehrbuch, 3rd edition, 2008.

[11] H. H. Binder. *Kleines Lexikon der chemischen Elemente*. Lehmanns Media, 2011.

[12] D. J. Ward. Future energy and the role for fusion. Presentation, July 2008. Summer School for Fusion Technologies, Forschungszentrum Karlsruhe GmbH, Karlsruhe, Germany.

[13] O. J. Weiß, E. Gaganidze, and J. Aktaa. Quantitative characterization of microstructural defects in up to 32 dpa neutron irradiated EUROFER97. *Journal of Nuclear Materials*, 426:52–58, 2012.

[14] O. J. Weiß. TEM investigations on unirradiated and neutron irradiated EUROFER97 and boron doped model alloys. TEM was performed at Institute for Applied Materials, Karlsruhe Institute of Technology, Karlsruhe, Germany, *unpublished*, 2009–2011.

[15] N. Baluc. Materials for fusion power reactors. *Plasma Physics and Controlled Fusion*, 48:B165–B177, 2006.

[16] U. Fischer, S. Simakov, U. v. Möllendorff, P. Pereslavtsev, and P. Wilson. Validation of activation calculations using the Intermediate Energy Activation File IEAF-2001. *Fusion Engineering and Design*, 69(1-4):485–489, 2003.

[17] P. Dubuisson, D. Gilbon, and J. L. Séran. Microstructural evolution of ferritic-martensitic steels irradiated in the fast breeder reactor Phénix. *Journal of Nuclear Materials*, 205:178–189, 1993.

[18] J. L. Seran. Deformation of austenitic Phénix claddings compared to F/M materials. In *IAEA Technical Commitee Meeting*, 2006.

[19] P. J. Maziasz. A perspective on present and future alloy development efforts on austenitic stainless steels for fusion application. *Journal of Nuclear Materials*, 133-134:134–140, 1985.

[20] E. A. Little. Void-swelling in irons and ferritic steels. *Journal of Nuclear Materials*, 87(1):11–24, 1979.

[21] E. A. Little and D. A. Stow. Void-swelling in irons and ferritic steels. *Journal of Nuclear Materials*, 87(1):25–39, 1979.

[22] E. Materna-Morris, A. Möslang, R. Rolli, and H.-C. Schneider. Effect of helium on tensile properties and microstructure in 9%Cr-WVTa-steel after neutron irradiation up to 15 dpa between 250 and 450 °C. *Journal of Nuclear Materials*, 386-388:422–425, 2009.

[23] E. Gaganidze, B. Dafferner, H. Ries, R. Rolli, H. C Schneider, and J. Aktaa. Irradiation Programme HFR Phase IIb - SPICE Impact testing on up to 16.3 dpa irradiated RAFM steels. *Wissenschaftliche Berichte, Forschungszentrum Karlsruhe*, FZKA 7371, 2008.

[24] E. Gaganidze, H.-C. Schneider, B. Dafferner, and J. Aktaa. Embrittlement behavior of neutron irradiated RAFM steels. *Journal of Nuclear Materials*, 367-370:81–85, 2007.

[25] H. Ullmaier and J. Chen. Low temperature tensile properties of steels containing high concentrations of helium. *Journal of Nuclear Materials*, 318:228–233, 2003.

[26] S. Fréchard, M. Walls, M. Kociak, J. P. Chevalier, J. Henry, and D. Gorse. Study by EELS of helium bubbles in a martensitic steel. *Journal of Nuclear Materials*, 393:102–107, 2009.

[27] I.-S. Kim, J. D. Hunn, N. Hashimoto, D. L. Larson, P. J. Maziasz, K. Miyahara, and E. H. Lee. Defect and void evolution in oxide dispersion strengthened ferritic steels under 3.2 mev fe+ ion irradiation with simultaneous helium injection. *Journal of Nuclear Materials*, 280(3):264–274, 2000.

[28] R. J. Kurtz, G. R. Odette, T. Yamamoto, D. S. Gelles, P. Miao, and B. M. Oliver. The transport and fate of helium in martensitic steels at fusion relevant He/dpa ratios and dpa rates. *Journal of Nuclear Materials*, 367-370(Part 1):417–422, 2007.

[29] E. Materna-Morris, A. Möslang, H.-C. Schneider, and R. Rolli. Microstructure and tensile properties in reduced activation 8-9% Cr steels at fusion relevant He/dpa ratios, dpa rates and irradiation temperatures. *Proc. of 22nd IAEA Fusion Energy Conference*, FT/P2-2:1–5, 2008.

[30] E. Materna-Morris, J. Aktaa, B. Dafferner, J. Ehrmann, E. Gaganidze, M. Holzer, S. Lautensack, A. Möslang, H. Ries, R. Rolli, and H.-C. Schneider. Tensile, charpy and fatigue specimen testing after neutron irradiation up to 15 dpa in the range of 250°C - 450°C, completion of the irradiation and post-irradiation examination. *Forschungszentrum Karlsruhe*, FZKA 7383:129–133, 2008.

[31] E. Gaganidze, C. Dethloff, O. J. Weiß, V. Svetukhin, M. Tikhonchev, and J. Aktaa. Modeling and TEM Investigation of Helium Bubble Growth in RAFM Steels under Neutron Irradiation. In *25th Symposium on Effects of*

Radiation on Nuclear Materials, 2011. Anaheim, CA, United States of America.

[32] Y. Dai and W. Wagner. Materials researches at the Paul Scherrer Institute for developing high power spallation targets. *Journal of Nuclear Materials*, 389(2):288–296, 2009.

[33] E. Wakai, S. Jitsukawa, H. Tomita, K. Furuya, M. Sato, K. Oka, T. Tanaka, F. Takada, T. Yamamoto, Y. Kato, Y. Tayama, K. Shiba, and S. Ohnuki. Radiation hardening and -embrittlement due to He production in F82H steel irradiated at 250°C in JMTR. *Journal of Nuclear Materials*, 343(1-3):285–296, 2005.

[34] Y. Dai, J. Henry, Z. Tong, X. Averty, J. Malaplate, and B. Long. Neutron/proton irradiation and He effects on the microstructure and mechanical properties of ferritic/martensitic steels T91 and EM10. *Journal of Nuclear Materials*, 415(3):306–310, 2011.

[35] C. Petersen, V. Shamardin, A. Fedoseev, G. Shimansky, V. Efimov, and J. W. Rensman. The ARBOR irradiation project. *Journal of Nuclear Materials*, 307-311:1655–1659, 2002.

[36] J. Ahlf, R. Conrad, G. P. Tartaglia, and G. Tsotridis. The HFR Petten as a test bed for fusion materials and components. *Journal of Nuclear Materials*, 212-215:1635–1639, 1994.

[37] R. R. Melder, V. P. Chakin, A. S. Pokrovsky, and A. N. Shchuchkin. The SSC RIAR high-flux research reactors: experience and possibilities of testing materials and mock-ups for fusion. *Fusion Engineering and Design*, 69(1-4):409–417, 2003.

[38] A. Möslang. IFMIF: the intense neutron source to qualify materials for fusion reactors. *Comptes Rendus Physique*, 9:457–468, 2008.

[39] N. Baluc, K. Abe, J. L. Boutard, V. M. Chernov, E. Diegele, S. Jitsukawa, A. Kimura, R. L. Klueh, A. Kohyama, R. J. Kurtz, R. Lässer, H. Matsui, A. Möslang, T. Muroga, G. R. Odette, M. Q. Tran, B. Schaaf, Y. Wu, J. Yu, and S. J. Zinkle. Status of R&D activities on materials for fusion power reactors. *Nuclear Fusion*, 47(10):S696–S717, 2007.

[40] R. L. Klueh, N. Hashimoto, M. A. Sokolov, P. J. Maziasz, K. Shiba, and S. Jitsukawa. Mechanical properties of neutron-irradiated nickel-containing martensitic steels: II. Review and analysis of helium-effects studies. *Journal of Nuclear Materials*, 357:169–182, 2006.

[41] P. Jung, J. Henry, J. Chen, and J. C Brachet. Effect of implanted helium on tensile properties and hardness of 9% Cr martensitic stainless steels. *Journal of Nuclear Materials*, 318:241–248, 2003.

[42] H. Trinkaus. The effect of cascade induced gas resolution on bubble formation in metals. *Journal of Nuclear Materials*, 318:234–240, 2003.

[43] R. J. Kurtz, A. Alamo, E. Lucon, Q. Huang, S. Jitsukawa, A. Kimura, R. L. Klueh, G. R. Odette, C. Petersen, M. A. Sokolov, P. Spätig, and J.-W. Rensman. Recent progress toward development of reduced activation ferritic/martensitic steels for fusion structural applications. *Journal of Nuclear Materials*, 386-388:411–417, 2009.

[44] Paul Scherrer Institute (Switzerland). The SINQ Facility. online. http://asq.web.psi.ch/facility/.

[45] N. Baluc, D. S. Gelles, S. Jitsukawa, A. Kimura, R. L. Klueh, G. R. Odette, B. Schaaf, and J. Yu. Status of reduced activation ferritic/martensitic steel development. *Journal of Nuclear Materials*, 367-370:33–41, 2007.

[46] B. D. Wirth, G. R. Odette, J. Marian, L. Ventelon, J. A. Young-Vandersall, and L. A. Zepeda-Ruiz. Multiscale modeling of radiation damage in Fe-based alloys in the fusion environment. *Journal of Nuclear Materials*, 329-333:103–111, 2004.

[47] D. R. Hartree. The Wave Mechanics of an Atom with a Non-Coulomb Central Field. Part I. Theory and Methods. *Mathematical Proceedings of the Cambridge Philosophical Society*, 24:89–110, 1928.

[48] V. Fock. Näherungsmethode zur Lösung des quantenmechanischen Mehrkörperproblems. *Zeitschrift für Physik A Hadrons and Nuclei*, 61(1):126–148, 1930.

[49] L. H. Thomas. The calculation of atomic fields. *Mathematical Proceedings of the Cambridge Philosophical Society*, 23(05):542–548, 1927.

[50] P. A. M. Dirac. Note on Exchange Phenomena in the Thomas Atom. *Mathematical Proceedings of the Cambridge Philosophical Society*, 26:376–385, 1930.

[51] P. Hohenberg and W. Kohn. Inhomogeneous Electron Gas. *Phys. Rev.*, 136(3B):B864–B871, 1964.

[52] W. Kohn and L. J. Sham. Self-Consistent Equations Including Exchange and Correlation Effects. *Phys. Rev.*, 140(4A):A1133–A1138, 1965.

[53] D. C. Rapaport. *The Art of Molecular Dynamics Simulation*. Cambridge University Press, 2004.

[54] M. Tikhonchev, V. Svetukhin, A. Kadochkin, and E. Gaganidze. MD simulation of atomic displacement cascades in Fe-10 at.%Cr binary alloy. *Journal of Nuclear Materials*, 395(1-3):50–57, 2009.

[55] U. Burghaus, J. Stephan, L. Vattuone, and J. M. Rogowska. *A Practical Guide to Kinetic Monte Carlo Simulations and Classical Molecular Dynamic Simulations*. Nova Science Publishers, 2006.

[56] G. V. Müller. *Monte Carlo Computersimulation mit* ab initio *Potentialen zur Berechnung thermodynamischer Größen*. VDI Verlag GmbH, 1998. Dissertation.

[57] N. M. Ghoniem and S. Sharafat. A numerical solution to the Fokker-Planck equation describing the evolution of the interstitial loop microstructure during irradiation. *Journal of Nuclear Materials*, 92(1):121–135, 1980.

[58] M. P. Surh, J. B. Sturgeon, and W. G. Wolfer. Master equation and Fokker-Planck methods for void nucleation and growth in irradiation swelling. *Journal of Nuclear Materials*, 325(1):44–52, 2004.

[59] F. F. Abraham. *Homogeneous Nucleation Theory - The Pretransition Theory of Vapor Condensation*. Academic Press, Inc., 1974.

[60] V. V. Slezov, J. W. P. Schmelzer, and A. S. Abyzov. *Nucleation Theory and Applications*. Wiley-VCH, 2005.

[61] H. Trinkaus and B. N. Singh. Helium accumulation in metals during irradiation - where do we stand? *Journal of Nuclear Materials*, 323:229–242, 2003.

[62] G. S. Was. *Fundamentals of Radiation Materials Science*. Springer-Verlag, 2007.

[63] R. E. Stoller and G. R. Odette. Analytical solutions for helium bubble and critical radius parameters using a hard sphere equation of state. *Journal of Nuclear Materials*, 131(2-3):118–125, 1985.

[64] L. K. Mansur, E. H. Lee, P. J. Maziasz, and A. P. Rowcliffe. Control of helium effects in irradiated materials based on theory and experiment. *Journal of Nuclear Materials*, 141-143:633–646, 1986.

[65] C. J. Ortiz, M. J. Caturla, C. C. Fu, and F. Willaime. He diffusion in irradiated alpha-Fe: An ab-initio-based rate theory model. *Physical Review B*, 75:100102–4, 2007.

[66] C.-C. Fu and F. Willaime. Ab initio study of helium in alpha-Fe: Dissolution, migration, and clustering with vacancies. *Physical Review B*, 72:064117–6, 2005.

[67] V. A. Borodin and P. V. Vladimirov. Kinetic properties of small He-vacancy clusters in iron. *Journal of Nuclear Materials*, 386-388:106–108, 2009.

[68] D. Terentyev, N. Juslin, K. Nordlund, and N. Sandberg. Fast three dimensional migration of He clusters in bcc Fe and Fe–Cr alloys. *Journal of Applied Physics*, 105:103509–12, 2009.

[69] L. K. Mansur. *Kinetics of Nonhomogeneous Processes - 8. Mechanisms and Kinetics of Radiation Effects in Metals and Alloys*. Wiley-Interscience Publication, 1987.

[70] K. Morishita. Nucleation path of helium bubbles in metals during irradiation. *Philosophical Magazine*, 87:1139–1158, 2007.

[71] S. M. Hafez Haghighat, G. Lucas, and R. Schäublin. State of pressurized helium bubble in iron. *EPL (Europhysics Letters)*, 85(6):60008, 2009.

[72] G. E. Lucas. The evolution of mechanical property change in irradiated austenitic stainless steels. *Journal of Nuclear Materials*, 206:287–305, 1993.

[73] F. A. Garner, M. L. Hamilton, N. F. Panayotou, and G. D. Johnson. The microstructural origins of yield strength changes in AISI 316 during fission or fusion irradiation. *Journal of Nuclear Materials*, 104:803–807, 1981.

[74] J. Gan and G. S. Was. Microstructure evolution in austenitic Fe-Cr-Ni alloys irradiated with protons: comparison with neutron-irradiated microstructures. *Journal of Nuclear Materials*, 297:161–175, 2001.

[75] N. Hashimoto, T. S. Byun, K. Farrell, and S. J. Zinkle. Deformation microstructure of neutron-irradiated pure polycrystalline vanadium. *Journal of Nuclear Materials*, 336:225–232, 2005.

[76] R. Lindau, A. Möslang, M. Rieth, M. Klimiankou, E. Materna-Morris, A. Alamo, A.-A. F. Tavassoli, C. Cayron, A.-M. Lancha, P. Fernandez, N. Baluc, R. Schäublin, E. Diegele, G. Filacchioni, J. W. Rensman, B. v.d. Schaaf, E. Lucon, and W. Dietz. Present development status of EUROFER and ODS-EUROFER for application in blanket concepts. *Fusion Engineering and Design*, 75-79:989–996, 2005.

[77] P. Graf, H. Zimmermann, E. Nold, E. Materna-Morris, and A. Möslang. Der Einfluss von Bor auf die Gefügeeigenschaften von martensitischen 9%-Chromstählen. *Sonderbände der Praktischen Metallographie*, 35:71–76, 2004.

[78] A. Möslang, E. Diegele, M. Klimiankou, R. Lässer, R. Lindau, E. Lucon, E. Materna-Morris, C. Petersen, R. Pippan, J. W. Rensman, M. Rieth, B. Schaaf, H.-C. Schneider, and F. Tavassoli. Towards reduced activation structural materials data for fusion DEMO reactors. *Nuclear Fusion*, 45(7):649–655, 2005.

[79] E. Gaganidze and J. Aktaa. The effects of helium on the embrittlement and hardening of boron doped EUROFER97 steels. *Fusion Engineering and Design*, 83:1498–1502, 2008.

[80] C. Petersen. Post irradiation examination of RAF/M steels after fast reactor irradiation up to 33 dpa and ¡340 °C (ARBOR 1). *Wissenschaftliche Berichte, Forschungszentrum Karlsruhe*, FZKA 7517:1–168, 2010.

[81] E. Gaganidze, C. Petersen, E. Materna-Morris, C. Dethloff, O. J. Weiß, J. Aktaa, A. Povstyanko, A. Fedoseev, O. Makarov, and V. Prokhorov. Mechanical properties and TEM examination of RAFM steels irradiated up to 70 dpa in BOR-60. *Journal of Nuclear Materials*, 417:93–98, 2011.

[82] E. Gaganidze and C. Petersen. Post irradiation examination of RAFM steels after fast reactor irradiation up to 71 dpa and ¡340 °C (ARBOR 2). *KIT Scientific Reports*, 7596:1–174, 2011.

[83] M. Rieth, B. Dafferner, and H. D. Röhrig. Charpy impact properties of low activation alloys for fusion applications after neutron irradiation. *Journal of Nuclear Materials*, 233–237(1):351–355, 1996.

[84] C. Dethloff, E. Nold, T. Weingärtner, and E. Gaganidze. Auger Electron Spectrometry (AES) analysis on boron distribution in unirradiated boron doped EUROFER97 model alloys. AES analysis performed at Institute for Applied Materials, Karlsruhe Institute of Technology, Karlsruhe, Germany, *unpublished*, 2008.

[85] L. L. Hsiung, M. J. Fluss, S. J. Tumey, B. W. Choi, Y. Serruys, F. Willaime, and A. Kimura. Formation mechanism and the role of nanoparticles in Fe-Cr ODS steels developed for radiation tolerance. *Phys. Rev. B*, 82(18):184103-1–13, 2010.

[86] Y. Serruys, M.-O. Ruault, P. Trocellier, S. Miro, A. Barbu, L. Boulanger, O. Kaitasov, S. Henry, O. Leseigneur, P. Trouslard, S. Pellegrino, and S. Vaubaillon. JANNUS: experimental validation at the scale of atomic modelling. *Comptes Rendus Physique*, 9(3-4):437–444, 2008.

[87] T. Yamamoto, G. R. Odette, P. Miao, D. J. Edwards, and R. J. Kurtz. Helium effects on microstructural evolution in tempered martensitic steels: In situ helium implanter studies in HFIR. *Journal of Nuclear Materials*, 386-388:338–341, 2009.

[88] M. Tikhonchev. Private communication, 2011. Ulyanovsk State University, Ulyanovsk, Russia.

[89] R. E. Stoller and D. M. Stewart. An atomistic study of helium resolution in bcc iron. *Journal of Nuclear Materials*, 417(1-3):1106–1109, 2011.

[90] R. E. Stoller. The Potential Impact of Ballistic Helium Resolutioning on Bubble Evolution. In *15th International Conference on Fusion Reactor Materials*, 2011. October 16th to 22nd, Charleston, SC, USA.

[91] G. J. Ackland, M. I. Mendelev, D. J. Srolovitz, S. Han, and A. V. Barashev. Development of an interatomic potential for phosphorus impurities in α-iron. *Journal of Physics: Condensed Matter*, 16(27):S2629–S2642, 2004.

[92] N. Juslin and K. Nordlund. Pair potential for Fe–He. *Journal of Nuclear Materials*, 382(2-3):143–146, 2008.

[93] D. E. Beck. A new interatomic potential function for helium. *Molecular Physics*, 14(4):311–315, 1968.

[94] S. V. Bulyarskii, V. V. Svetukhin, and O. V. Prikhod'ko. Atomic structure and non-electronic properties of semiconductor - Spatially inhomogeneous oxygen precipitation in silicon. *Semiconductors*, 33:1157–1162, 1999.

[95] P. Pichler. *Intrinsic Point Defects, Impurities, and Their Diffusion in Silicon*. Computational Microelectronics. Springer-Verlag Wien, 2004.

[96] C. J. Ortiz, P. Pichler, T. Fuhner, F. Cristiano, B. Colombeau, N. E. B. Cowern, and A. Claverie. A physically based model for the spatial and temporal evolution of self-interstitial agglomerates in ion-implanted silicon. *Journal of Applied Physics*, 96:4866–4877, 2004.

[97] R. T. DeHoff. *Thermodynamics in Materials Science*. CRC Press, 2nd edition, 2006.

[98] A. Gokhman, F. Bergner, A. Ulbricht, and U. Birkenheuer. Cluster dynamics simulation of reactor pressure vessel steels under irradiation. *Defect and Diffusion Forum*, 277:75–80, 2008.

[99] A. Gokhman, F. Bergner, and A. Ulbricht. Modelling of vacancy cluster evolution in neutron irradiated iron. *FZ-Dresden*, 04-R13:76–81, 2009.

[100] W. R. Tyson and W. A. Miller. Surface free energies of solid metals: Estimation from liquid surface tension measurements. *Surface Science*, 62:267–276, 1977.

[101] R. Kemp, G. Cottrell, and K. D. H. Bhadeshia. Classical thermodynamic approach to void nucleation in irradiated materials. *Energy Materials: Materials Science and Engineering for Energy Systems*, 1(2):103–105, 2006.

[102] H. Ullmaier. Landolt-Börnstein - Group III Condensed Matter - Numerical Data and Functional Relationships in Science and Technology. *The Landolt-Börnstein Database*, 25: Atomic Defects in Metals:380–437, 2010.

[103] L. Yang, X. T. Zu, H. Y. Xiao, F. Gao, H. L. Heinisch, and R. J. Kurtz. Defect production and formation of helium-vacancy clusters due to cascades in α-iron. *Physica B: Condensed Matter*, 391(1):179–185, 2007.

[104] P. G. Klemens and B. Cort. Thermodynamic properties of helium bubbles in aged plutonium. *Journal of Alloys and Compounds*, 252:157–161, 1997.

[105] G. R. Odette, M. J. Alinger, and B. D. Wirth. Recent developments in irradiation-resistant steels. *Annual Review of Materials Research*, 38:471–503, 2008.

[106] F. A. Nichols. On the estimation of sink-absorption terms in reaction-rate-theory analysis of radiation damage. *Journal of Nuclear Materials*, 75:32–41, 1978.

[107] A. D. Brailsford and R. Bullough. The theory of sink strengths. *Philosophical Transactions of the Royal Society of London. Series A, Mathematical and Physical Sciences (1934-1990)*, 302:87–137, 1981.

[108] F. Gao, H. L. Heinisch, and R. J. Kurtz. Migration of vacancies, He interstitials and He-vacancy clusters at grain boundaries in α-Fe. *Journal of Nuclear Materials*, 386-388:390–394, 2009.

[109] I. D. Chivers and J. Sleightholme. *Introduction to Programming with Fortran - With Coverage of Fortran 90, 95, 2003, and 77*. Springer-Verlag London, 2006.

[110] The HSL Mathematical Software Library. online, 2011. http://www.hsl.rl.ac.uk/.

[111] L. Powers and M. Snell. *Microsoft Visual Studio 2008 Unleashed*. SAMS Pearson Education, Inc., 2008.

[112] Intel Software Network. online, 2011. http://software.intel.com/en-us/articles/intel-compilers/.

[113] B. Chapman, G. Jost, and R. van der Pas. *Using OpenMP - Portable Shared Memory Parallel Programming*. Scientific and Engineering Computation. The MIT Press, 2008.

[114] J. Crank. *The Mathematics of Diffusion*. Oxford University Press Inc., New York, 2nd edition, 1975.

[115] E. Hairer, S. P. Norsett, and G. Wanner. *Solving Ordinary Differential Equations I – Nonstiff Problems*. Springer Series in Computational Mathematics 8. Springer-Verlag, 2nd edition, 1993.

[116] M. Rieth, B. Dafferner, and H. D. Röhrig. Embrittlement behaviour of different international low activation alloys after neutron irradiation. *Journal of Nuclear Materials*, 258-263:1147–1152, 1998.

[117] L. R. Greenwood, B. M. Oliver, S. Ohnuki, K. Shiba, Y. Kohno, A. Kohyama, J. P. Robertson, J. W. Meadows, and D. S. Gelles. Accelerated helium and hydrogen production in 54Fe doped alloys - measurements and calculations for the FIST experiment. *Journal of Nuclear Materials*, 283-287:1438–1442, 2000.

[118] G. Shimansky. Private communication, 2007. SSC RIAR, Dimitrovgrad, Russia.

[119] G. Speich, A. Schwoeble, and W. Leslie. Elastic constants of binary iron-base alloys. *Metallurgical and Materials Transactions B*, 3:2031–2037, 1972.

[120] O. J. Weiß. Private communication, 2010. Karlsruhe Institute of Technology, Institute for Materials Research II, Karlsruhe, Germany.

[121] C. Dethloff, E. Gaganidze, V. V. Svetukhin, and J. Aktaa. Modeling of helium bubble nucleation and growth in neutron irradiated boron doped RAFM steels. *Journal of Nuclear Materials*, 426:287–297, 2012.

[122] A. Caro, J. Hetherly, A. Stukowski, M. Caro, E. Martinez, S. Srivilliputhur, L. Zepeda-Ruiz, and M. Nastasi. Properties of Helium bubbles in Fe and FeCr alloys. *Journal of Nuclear Materials*, 418(1-3):261–268, 2011.

[123] D. A. Terentyev, L. Malerba, R. Chakarova, K. Nordlund, P. Olsson, M. Rieth, and J. Wallenius. Displacement cascades in Fe-Cr: A molecular dynamics study. *Journal of Nuclear Materials*, 349(1-2):119–132, 2006.

[124] L. R. Greenwood and R. K. Smither. SPECTER: Neutron Damage Calculations for Materials Irradiations. Technical Report ANL/FPP/TM-197, Argonne National Laboratory, IL, USA, 1985.

[125] E. Daum and U. Fischer. Long-term activation potential of the steel Eurofer as structural material of a DEMO breeder blanket. *Fusion Engineering and Design*, 49–50:529–533, 2000.

[126] C. Dethloff, E. Gaganidze, V. Svetukhin, and J. Aktaa. Modeling of Helium Bubble Growth in 10B-doped RAFM Steels under Neutron Irradiation. In *15th International Conference on Fusion Reactor Materials*, 2011. October 16th to 22nd, Charleston, SC, USA.

[127] M. Lambrecht, E. Meslin, L. Malerba, M. Hernández-Mayoral, F. Bergner, P. Pareige, B. Radiguet, and A. Almazouzi. On the correlation between irradiation-induced microstructural features and the hardening of reactor pressure vessel steels. *Journal of Nuclear Materials*, 406(1):84–89, 2010.

Schriftenreihe
des Instituts für Angewandte Materialien

ISSN 2192-9963

Die Bände sind unter www.ksp.kit.edu als PDF frei verfügbar oder
als Druckausgabe bestellbar.

Band 1   Prachai Norajitra
          Divertor Development for a Future Fusion Power Plant. 2011
          ISBN 978-3-86644-738-7

Band 2   Jürgen Prokop
          **Entwicklung von Spritzgießsonderverfahren zur Herstellung von
          Mikrobauteilen durch galvanische Replikation.** 2011
          ISBN 978-3-86644-755-4

Band 3   Theo Fett
          **New contributions to R-curves and bridging stresses – Applications
          of weight functions.** 2012
          ISBN 978-3-86644-836-0

Band 4   Jérôme Acker
          **Einfluss des Alkali/Niob-Verhältnisses und der Kupferdotierung auf
          das Sinterverhalten, die Strukturbildung und die Mikrostruktur von
          bleifreier Piezokeramik ($K_{0,5}Na_{0,5}$)NbO_3.** 2012
          ISBN 978-3-86644-867-4

Band 5   Holger Schwaab
          **Nichtlineare Modellierung von Ferroelektrika unter Berücksich-
          tigung der elektrischen Leitfähigkeit.** 2012
          ISBN 978-3-86644-869-8

Band 6   Christian Dethloff
          **Modeling of Helium Bubble Nucleation and Growth in Neutron
          Irradiated RAFM Steels.** 2012
          ISBN 978-3-86644-901-5